5G and Satellite Spectrum, Standards, and Scale

For a complete listing of titles in the
Artech House Space Technology and Applications Series,
turn to the back of this book.

5G and Satellite Spectrum, Standards, and Scale

Geoff Varrall

ARTECH HOUSE
BOSTON | LONDON
artechhouse.com

Library of Congress Cataloging-in-Publication Data
A catalog record for this book is available from the U.S. Library of Congress.

British Library Cataloguing in Publication Data
A catalogue record for this book is available from the British Library.

Cover design by John Gomes

ISBN 13: 978-1-63081-502-8

685 Canton Street
Norwood, MA 02062

10 9 8 7 6 5 4 3 2 1

Contents

Foreword

We live in an always-on world, where our social, political, and working lives are driven and supported by connectivity in a way that would have been scarcely believable 20 years ago. A huge proliferation of smart devices, the majority being machine-to-machine devices, is quietly and efficiently automating the functioning of our world, fed by an explosion of cloud-based applications. These two forces are brought together and empowered by connectivity, especially mobile connectivity. With 5G and the Internet of Things (IoT) looming on the horizon, these trends are set to accelerate dramatically.

Every aspect of the world we live in is being profoundly changed by these forces. In the years to come, we will live in smart cities, travel to work across intelligent and autonomous transport systems, pass through smart borders, maintain our health and well-being through wearable technology, and live in a much greener and safer environment thanks to smart agriculture and fishing, aviation, and merchant marine activities.

To support and enable this exciting new world, rich, ubiquitous, and highly reliable connectivity will be essential; indeed, the negative impact on human potential of being without such connectivity will become so fundamental that connectivity will come to be seen as a basic human right. Conversely, with connectivity the digital society becomes a truly global phenomenon, binding our planet together for mutual benefit. Yet today, more than 4 billion humans live their lives without access to the internet: this digital divide is one of the big challenges of our age.

In this context, the emergence of next-generation space-based capabilities offers a truly exciting potential to support, enable, and extend this digital society. We are living through a golden age of space-based innovation, at a time when such innovation has never been more important for human development.

The delivery from space of ubiquitous, highly reliable, and cost-effective connectivity, broadcast services, Earth observation capabilities, and precision location services offers us the chance not only to close the global digital divide in the developing world but also to enhance the emerging digital society in the developed world. The satellite industry's unique capabilities will extend the digital society into remote areas, onto the seas and into the skies, and ensure absolute security and reliability of the networks that will increasingly run our world in the twenty-first century.

In the 5G context, this means that space-based capabilities will become a key component of 5G deployments: a vital contributor to heterogeneous networks that co-opt many different complementary technologies to deliver on the promise of 5G to the society that it intends to serve. As such, this book offers an important new perspective, postulating the need for regulators, standards-setting bodies, and market participants to come together to support the inclusion of space-based capabilities into the 5G firmament, indeed, as an essential driver of the future success of 5G networks globally. I commend this important work to readers.

Rupert Pearce, CEO, Inmarsat
May 2018

Acknowledgments

This book is written as a direct follow-up to our earlier book, *5G Spectrum and Standards*, published by Artech House in 2016. The original book is available from Artech House and is useful although not essential to read prior to engaging with this our latest effort. *5G Spectrum and Standards* reviews the spectrum band plan outcomes of 2015 World Radio Congress (WRC) and the Third Generation Partnership Project (3GPP) 5G standards process covering Releases 15 and 16 as documented 2 years ago. It also reviews the present users of the centimeter band between 3 GHz and 30 GHz and the millimeter band between 30 and 300 GHz including near space and deep space communication and observation systems and outlines some of the emerging coexistence issues implied by an increased cosharing of this spectrum. The book includes useful contributions from Sylvia Lu of u-blox on the practical constraints of digital signal processor (DSP) bandwidth on device power budgets.

This new book, *5G and Satellite Spectrum, Standards, and Scale*, brings us up to date with 3GPP Releases 16 and 17 and the related New Radio physical layer specifications. It also captures the emerging focus on discrete vertical markets and their specific physical layer and upper layer protocol and performance requirements.

However, over the past 2 years, we have observed a remarkable technical and commercial transformation in the satellite industry with new service models emerging both from existing established geostationary orbit (GSO), medium Earth orbit (MEO), and low Earth orbit (LEO) operators, and from NEWLEO operators such as OneWeb, Space X, and LeoSat.

In essence, a combination of hardware innovation, manufacturing innovation, launch innovation, constellation innovation, and business model innovation is having a profound impact on delivery cost and performance both at individual user level and IoT device level and a persuasive case can be made that many 5G vertical market use cases could potentially be served more efficiently from space.

These satellite systems scale from ultrahigh frequency (UHF) though L-band, S-band, and C-band with broadband fixed and mobile wireless connectivity delivered primarily over a mix of Ku-, K-, and Ka-band spectral assets. Cell radii scale from 2 km to 2,000 km.

Coexistence between NEWLEO and established GSO and MEO and LEO operators is managed by implementing a range of angular power separation techniques combined with power control algorithms that potentially support frequency use both between multiple satellite constellations and 5G terrestrial services but many regulatory and competitive positioning issues need to be resolved before these techniques can be deployed and trusted universally.

5G and Satellite Spectrum, Standards, and Scale documents these techniques and the associated interference modeling and explores the evolving business and financial and commercial implications of this transformation process. In particular, we study the implications for WRC 2019 (the follow up World Radio Congress to WRC2015) and related 5G and satellite standards work and in broader terms discuss the likely impact on the future enterprise value of satellite and 5G operators and their associated supply chains and the role of other stakeholders including Google, Apple, Facebook, and Amazon and their emerging Asian competition (Alibaba and Tencent).

I am aided in this probably rather overambitious endeavour by two useful industry friends. Mr. Martin Sims and his research team at Policy Tracker added their regulatory insight at regular intervals throughout the text and Mr. John Tysoe at The Mobile World provided financial metrics from his remarkable database of operator and supply chain company filings, sharing with us gearing ratio comparisons that are at times startling but add significantly to our understanding of the financial dynamics of the telecommunications industry.

I would also like to thank my codirector, Roger Belcher, who over the past 30 years has patiently corrected my occasionally embarrassing lack of technical knowledge in the more arcane areas of radio frequency (RF) theory and practice. He has not been involved in this book due to his ongoing involvement in motorcycle racing, which he assures me is a proven antidote to aging. He therefore cannot be held responsible for any technical errors.

Also, thanks are due to Stirling Essex and the customers with whom we have worked on 5G and satellite vertical market business modeling over the past 2 years.

Finally, I thank my wife Liz who remains perplexed by my recurring need to write books on telecommunications (this is the sixth) but who nevertheless remains critically supportive.

1

Sixty Years of Satellites

1.1 Beginning with the Beach Ball

On October 4, 1957, the fortieth anniversary of the Bolshevik revolution, the former USSR launched Sputnik 1, the first artificial satellite in space. The size of a beach ball and weighing 83.6 kg, Sputnik had transmitters at 20.005 MHz (15m wavelength) and 40.002 MHz (7.5m wavelength). Sputnik is still in orbit, although it is not doing very much.

Sixty years on, Elon Musk, founder of Space X and the Tesla Motor Company assures us that we will be soon be living on Mars [1] and flying anywhere on Earth in under an hour [2], Jeff Bezos of Amazon [3], and Richard Branson of Virgin Atlantic [4] have plans for us to vacation in space and Mark Zuckerberg of Facebook aims to connect the unconnected from space [5]. Not to be outdone, The Alphabet Group, the parent holding company for Google, and the investment group Fidelity have invested $1 billion in Space X in return for a 10% shareholding [6].

In parallel, the asteroid mining start-up Planetary Resources [7] has teamed up with the Duchy of Luxembourg to define a regulatory and legal framework for the ownership of mined resources from the asteroid belt. Goldman Sachs considers that the falling cost of rockets and the vast quantities of platinum sitting on space rocks makes this a hot investment prospect, though possibly better suited to the orphans' fund rather than the widows' fund [8].

Can a handful of new space entrepreneurs, relatively new companies (15 years ago Google had less than 12 employees) and one of the world's smallest but richest sovereign countries change an industry? Mr. Ford certainly made a big difference to the automotive industry and Mr. Marconi, in many ways an Edwardian version of Mr. Musk, made some big waves, or more precisely, long waves in the wireless industry.

It could be argued that the Marconi business empire was a product of the fading British empire, fueled by a mix of consumer and military spending. This model remains relevant today. Every time Kim Jong-un launches another ever longer-range missile over Japan, the U.S. ballistics budget gets bigger. Archimedes would have been surprised but probably pleased [9]. The principle of using a perceived enemy (for North Korea, the United States), as the justification for absolute control based on disproportionate military spending is well established. For Henry VIII, the threat from France was used to justify military spending that more or less bankrupted Tudor Britain but also helped to consolidate Henry's hold on absolute power. If Henry and Mr. Kim could meet today, they would have a lot in common, and Mr. Kim would undoubtedly be impressed by Henry's innovative financial remodeling of the medieval monasteries.

1.2 Russia, China, and the United States: Red Rockets and Yellow Rockets

This takes us back (or rather indirectly forward) to March 23, 1983, and to an Address to the Nation speech by President Ronald Reagan, which came to be known as his Star Wars speech (it coincided with *Return of the Jedi*, the third of the *Star Wars* films). The speech set out the rationale for an increase in defense spending on space-based missile interception predicated on the threat from Russia, the axis of evil as represented to the American public by the U.S. political and popular press. The impact of this shift in spending is still evident today, with Space X being active as a launch vehicle for the Boeing built X37B [10].

The developing military and commercial importance of space was recognized in April 2016 when Congressman Jim Bridenstein, the Republican representative of Oklahoma's First Congressional District, sponsored the U.S. Space Renaissance Act [11]. The Act describes space as the ultimate high ground and argues the case for more intensive use by the military of civilian satellite systems both for imaging and reconnaissance, attack detection, and space-based interception.

Space is also considered as crucial to future cybersecurity, though China (rather than the United States) has being making recent headlines with its successful distribution of quantum cryptographic keys from the Micius low Earth orbit (LEO) satellite achieving a distance of 1,200 km, 10 times the distance achieved to date over terrestrial fiber [12].

1.3 Space Regulation and Deregulation

Sixty years ago, Sputnik spurred the formation of NASA. The Cuban missile crisis of 1962 highlighted the strategic importance of space. The 1962 Satellite Communications Act "allowed the U.S. Government to supervise fair access for commercial satellites" and coincided with the launch of Telstar 1, the world's first communications satellite, followed in 1963 by the first geosynchronous satellite. The Satellite Communications Act created Comsat, which in 1964 became Intelsat with a membership of 17 nations. In April 1965, the first Intelsat satellite, Early Bird, was launched into geostationary orbit to deliver "TV and telephone and telegraph and high speed data," the world's first quad play platform. The Intelsat regulatory model was adopted in other regions. Eutelsat was formed in 1977 to operate the first European satellite (launched in 1983). Arabsat was founded in 1976 by the 21 member states of the Arab League.

Inmarsat (the International Maritime Satellite Organization) had a different starting point, set up as an international service operator in 1976 to oversee safety of life at sea (SOLAS). In 1982, Inmarsat started to provide mobile satellite communication services extending to land mobile in 1989 and aeronautical services in 1990. In 1999, Inmarsat was the first of the international satellite operators to deregulate as a response to the International Telecommunications Union (ITU) open skies policy. Intelsat and Eutelsat followed in 2001.

This was not good timing. The dotcom bubble had burst in 2000 and the telecom industry followed 2 years later. The dotcom boom had produced a feverish investment in transatlantic fiber and oversupply. All that unlit dark fiber meant that per-bit, long-distance delivery prices reduced to almost zero. In parallel, the satellite operators needed to maintain existing terrestrial and space hardware and put together plausible investment plans for new Ku-band, K-band, and Ka-band constellations. The result was that the satellite sector started to run uncomfortably high debt ratios. The debt servicing cost of Intelsat is presently equivalent to buying three satellites a year.

Fortuitously, income from TV including income from fully amortized C-band satellites and military payloads have helped to save the day. If Intelsat is excluded from a financial analysis of satellite operators, the sector is not currently overgeared, but it is a tribute to the satellite industry and their patient shareholders that they survived their first 15 years in the private sector and remain in a position to justify new research and development and hardware and software investment.

1.4 The Beach in Bournemouth

To get a real flavor of the potential of new space, we need to take a visit to the beach in Bournemouth. Imagine you are a flat-panel phased array antenna sitting in a deckchair staring into space. Depending on your latitude, you will have radio frequency (RF) visibility to at least 50 satellites and this is before 10,000 new low Earth orbit satellites (NEWLEOs) arrive in orbit. The smartphone by your side will have RF visibility to at most six cellular sites. It takes 20 minutes for a LEO to travel into space, significantly faster than a truck drop to a cellular site. Having unfurled its antennas, the LEO is ready to go and depending on how it is configured can stay in space for up to 20 years. Outside the Earth's atmosphere, solar energy density is 1,350 W/m^2. At the Earth's surface, it is 1,000 W/m^2. It is sunnier in space. It does not rain in space. Multijunction solar panel cells are now achieving 40% efficiency, so that is 20 years of free RF power and no rent to pay. Network densification is also easier (less expensive) in space. (There is more space in space.) It is also cold in space (–270.45°C), so there is no air conditioning to worry about.

If I want to do some high-frequency trading from my deck chair, I can get to the other side of the world significantly faster over an intersatellite-switched LEO constellation. Radio waves and light travel faster in free space than in a fiber optic cable. Once a fiber optic cable reaches a certain length (about 10,000 km), the free-space speed advantage outweighs the round trip distance (1,400 km).

Bournemouth, a popular U.K. south coast resort, happens to be one of the towns in the United Kingdom with the worst 4G coverage [13]. From my Bournemouth Council-supplied deck chair, I can get to Singapore via a LEO satellite network in 120 ms, which is at least 60 ms faster than fiber. LeoSat are basing their LEO business model on this time differential. If I really wanted to speed things up, then the transaction server would not be in Singapore but in the constellation (with interesting tax implications, another opportunity for those hotshot Luxembourg lawyers).

By contrast, if I used my smartphone, my journey to Singapore will be via the local 4G or 5G network; across a microwave link or fiber, cable, or copper backhaul; then to Singapore, which could be along a number of possible routes; and then into a Singapore network and finally into the Singapore server.

This highlights two points. I have no visibility to the end-to-end delay across multiple 4G and 5G mobile broadband and backhaul networks. Additionally, I have no control over the latency variability (also known as jitter). Apart from introducing uncertainty into the timing of the trade, it also makes authentication harder to manage. Challenge and response algorithms depend on deterministic round-trip latency and minimal jitter. In comparison, my end-to-end journey over the LEO constellation gives me absolute control of the end-to-end channel.

However, I forgot to mention that my deck chair has wheels and an electric motor. My LEO-based server tells me that it is sunnier and less crowded at the other end of the beach. I now have two choices. I can self-navigate myself along the beach using the dead reckoning (enabled by the real-time, high-accuracy clock pulse coming down from my nearest LEO satellite), or I can let the LEO drive me. It is probably easier to let the LEO take charge as it knows where all the other deck chairs are and knows that my battery is about to go flat so it can take me to the beach hut recharging point where I can take on some power and the latest software upgrade and buy some suntan lotion, a sun hat, and an ice cream. Bournemouth, by the way, claims to be one of the sunniest towns in Britain [14], but everything is relative.

1.5 Satellites for Autonomous Transport Systems and the Internet of Moving Objects

This is a trivial example but probably explains why Mr. Musk is keen to launch his own LEO satellite network. It will be extremely hard to deliver a totally safe semiautonomous or fully autonomous driving or terrestrial travel experience over multiple terrestrial cellular networks. It will be relatively easy to deliver a totally safe semiautonomous or fully autonomous driving or public transport experience over a LEO network. Mr. Musk may also have plans to conquer the mobile deckchair market, possibly another $50 billion opportunity.

However, this highlights a more general point. Server bandwidth on its own does not confer added value. The value comes from the control that accrues from the data held on the server and the algorithms used to mine and manage that data. This is a blindingly obvious statement but explains why the cloud comes (apparently) for free.

There are many stationary and moving objects that are already monitored and managed from space. Inmarsat and other operators (Iridium and Asia SAT) supply connectivity and management and monitoring systems to commercial aircraft. If my deck chair was on a Royal Caribbean cruise ship, it would be connected to the internet via the O3b MEO [15] constellation now owned and operated by SES. The constellation is also helping to ensure that the cruise ship does not crash into other cruise ships heading towards Bournemouth (O3b provides complementary support to the Maritime Automatic Identification System). Caterpillar, John Deere, Komatsu, and those other manufacturers of massive machines that dig very large holes and crop the wheat fields of America are shipped with Orbcomm very high frequency (VHF) modems for asset tracking and (low bandwidth) telemetry and telecommand.

We are describing an expansion of services that are already well established. Inmarsat started providing mobile satellite service in 1982 and a terrestrial service in 1989. Iridium, Globalstar, and Orbcomm have been providing mobile connectivity for 20 years, but these legacy services are based on two-way voice and data transmission rather than cloud connectivity.

1.6 Satellites and 5G: A Natural Convergence?

The combination of more satellites and more bandwidth and more onboard processing power and storage bandwidth significantly changes the market positioning of the satellite industry and brings it closer to emerging 5G business models.

OneWeb states that it is confident that it can substantially reduce 5G backhaul costs both in dense urban and deep rural areas and provide more cost-effective mobile and fixed broadband geographic coverage for rural connectivity [16]. This includes IoT connectivity and developing market connectivity where base station electricity is particularly expensive. In developed markets, the proposition could be particularly persuasive for operators presently overdependent on fiber owned and managed by their competitors.

The premise of this book is therefore simple. A range of technical, commercial, and regulatory innovations in the satellite industry are changing the delivery economics of space-based communication. This is sometimes described in the technical and commercial literature as new space or Space 2.0 (a reworking of Web 2.0).

This includes hardware innovation in space and on the ground, manufacturing innovation, launch innovation, and constellation innovation, in particular the development of mixed constellation delivery platforms combining the benefits of GSO, MEO, and LEO satellites. Constellation innovation includes techniques that allow the same passbands to be shared between constellations but also significantly with terrestrial 5G systems.

In the satellite industry, business models are based on a combination of spectral assets that include specific access rights to downlink and uplink spectrum, orbit rights, and what are usually called landing rights, the right to provide service into and out of sovereign nations visible from geostationary satellites or overflown by MEO and LEO satellites. An established customer base is also a prime asset.

In the 5G industry, business models are based on spectral access rights combined with picocell, microcell, and macrocell real estate and fiber and microwave backhaul. Money is borrowed on the basis that these access rights will be available over a known period, for example, 20 or 25 years or in some cases indefinitely provided that service obligations are achieved. As with the satellite industry, customers including IoT device subscriptions are an asset against which money can be borrowed and against which enterprise value is assessed.

For the past 30 years, the cellular and satellite industry have worked together on a modest scale. Approximately 1% of cellular network backhaul is carried over geostationary satellites. In some extreme geographic locations, satellites are the only way to connect a base station or are more economic than microwave or fiber or copper.

A new generation of satellite operators, to whom (for the sake of simplicity) we will refer to as NEWLEOs operators, aim to radically change this relationship.

1.7 The NEWLEOs

NEWLEO operators include OneWeb, Space X, and LeoSat. OneWeb and Space X have implementation plans based on launching hundreds and ultimately thousands of satellites into LEO. These high-count constellations use several gigahertz of uplink and downlink Ku-band, K-band, and Ka-band spectrum and longer-term plans to use V-band and W-band spectrum. The combination of this spectral bandwidth combined with superefficient solar panel arrays delivers sufficient RF power and capacity to support millions and potentially billions of users and devices on the ground both in terms of direct connectivity and backhaul provision.

This only makes sense if this connectivity can be delivered at equivalent or preferably lower cost than other options. NEWLEO investor presentations and regulatory filings are predicated on the assumption that delivery costs can be reduced to the point at which the presently disconnected, which apparently totals 35 million people in the United States and 3 to 4 billion people worldwide, can be connected cost-effectively.

Quite what this means is open to debate. For many of the presently disconnected living on a dollar or less a day, the notion of owning an Apple iPhone 10 at $1,000 remains a remote possibility. However, the costs reduce assuming that Wi-Fi can be used from a low-cost, solar-powered cell site serviced from a NEWLEO constellation. Additionally, the NEWLEO can argue that the subsidies presently going into rural fiber rollout could be spent more effectively on space-based systems, which presently receive less than 1.5% of government subsidy budgets on a global basis [17].

There are also potential performance gains in terms of long-distance latency. Iridium has successfully deployed a low-count LEO constellation (66 satellites), which has been providing service now for over 20 years with an ongoing constellation upgrade now in process. The constellation uses intersatellite switching in K-band between 23.187 GHz and 23.387 GHz.

Intersatellite switching has the benefit of reducing the number of Earth gateways needed but also provides absolute control of the end-to-end channel with reduced latency and minimal and known latency variability (also known as latency jitter). This makes Iridium well suited to a number of higher added-value military and safety-critical payloads.

LeoSat has a similar constellation proposal to Iridium based on the same space system platform provided by Thales but utilizing 7 GHz of paired spectrum (3.5 + 3.5 GHz) at Ka-band for individual user uplinks and downlinks (compared to 10 + 10 MHz of paired spectrum in L-band available to Iridium) and optical intersatellite switching. The U.S. Federal Communications Commission (FCC) filing is based on 120 to 140 satellites in a similar polar orbit to the Iridium Next Constellation. However, the business model is focused on providing a latency gain for high-value applications such as high-frequency trading, the oil and gas industries, and corporate networking, and government agencies (see Chapter 3 for more details). LeoSat is working with the European Space Agency on 5G and satellite transversal activities [18].

Similarly, Space X is proposing intersatellite switching using optical transceivers that would deliver similar latency gains. These could be uniquely useful in a number of global vertical markets including, for example, automotive connectivity and autonomous and semiautonomous cars, trucks, and transport systems.

Intersatellite switching can also be combined with interconstellation switching to provide additional cost savings. For example, LEO satellites can uplink to a GSO and then back to a GSO Earth gateway. This introduces additional latency but reduces the number of Earth stations. Given that a high-count LEO constellation could potentially require 50 gateways and that each gateway could have a capital cost of tens of millions of dollars and ongoing operational costs, then it can be seen that the potential savings are substantial.

1.8 Regulatory and Competition Policy

This brings us to related issues of regulatory policy and competition policy and operator competitive positioning. The established GSO operators have been working in some cases for over 50 years to consolidate their regulatory position both in terms of spectral assets, orbital rights, and landing rights, which includes the right to own and operate Earth gateways.

Low Earth satellites conveniently and inconveniently fly through the Earth-to-space and space-to-Earth paths of GSO and MEO satellites and potentially pour unwanted RF energy into satellite dishes on Earth pointing upwards at the same bit of sky.

This is not a problem for Iridium and Globalstar, both of whom have operated LEO constellations for 20 years because they have user links in L-band and, in the case of Iridium, military payloads, which justify spectral access priority.

The NEWLEO operators are, by contrast, deploying in Ku-band, K-band, and Ka-band in either the same passband as GSO operators or in adjacent spectrum. The NEWLEO operators are required to provide detailed evidence that sufficient mitigation measures are in place to meet the agreed protection ratios awarded to existing users of the spectrum.

This is achieved through angular power separation and power control mechanisms, which we cover in later chapters. However, the modeling used in these submissions is open to technical and legal challenge, particularly when multiple high-count NEWLEO operators or potential operators need to be accommodated. NEWLEO operators may also question each other's modeling methods, which weakens their position in relation to incumbent MEO and GSO operators.

In terms of commercial tension, the NEWLEO business models are predicated on rapid price declines based on assumed and projected rapid cost declines. By contrast, the GSO business models and MEO business models (O3b being one example) are based on relatively high price points with margins that provide adequate but not always generous cover for debt financing.

One solution would be for one or more of the NEWLEO entities to merge with one or more of the GSO and MEO operators. This could be technically compelling, but existing GSO and MEO operator bond holders need to be persuaded that higher gearing and increased implementation risk is worthwhile.

There may also be a nagging doubt that a merged entity could find that their spectral access rights, orbit rights, and landing rights open to legal challenge, which would be an alarming prospect. Some combination of these concerns is probably the explanation for the failed merger between OneWeb and Intelsat.

1.9 A Summary of Orbit Options and Performance Comparisons

Just as a reminder, it may be useful just to recap on the differences between LEO, MEO, and GSO. Satellite orbits can be categorized as shown in Table 1.1.

LEO constellations are normally deployed as polar orbits with the option to deploy satellites that are Sun-synchronous. Sun-synchronous satellites follow the dawn as it moves around the world. Sun-synchronous polar orbits are particularly effective for Earth imaging from space.

For the sake of completeness, we should also reference highly elliptical orbits (HEOs) such as the Tundra and Molnya orbits [19], although these orbits are best suited to high latitude and polar coverage and Quazi zenith constellations where some of the satellites are geosynchronous but not geostationary; a Global Navigation Satellite System (GNSS) backup constellation over Japan is one example [20].

Table 1.1
Satellites with Altitude (and Attitude)

LEO	160–2,000 km	99–1,200 miles
MEO	2,000–20,000 km	1,200–12,000 miles
GEO	36,000 km	22,000 miles

LEO, MEO, and GSO are the options of most interest to us in this book. In Figure 1.1, Inmarsat provided this nice graphic comparing the characteristics of the three orbit options including typical latencies and orbit duration.

The three orbital categories are generally used for different purposes. GSOs are aligned with the equator, and satellites in these orbits appear to be suspended motionless above a point on the Earth. These orbits are therefore useful for providing TV coverage and for weather observation. Spot beam antennas on the satellites can be used to provide coverage over specific areas of land or sea. Communications satellites in these orbits have a high path loss relative to MEO and LEO satellites and a longer round-trip delay. The additional path loss is accommodated by using high gain antennas. For example, very small aperture terminals (VSATs) have been used over the past 15 years to deliver high data rates to corporate and business and high-value personal users. An ongoing point in this book is that VSAT antennas are becoming more efficient both in the way that they deliver selective gain and reject unwanted signal energy. This gain in efficiency translates into lower delivery costs but, we also argue, helps to resolve many of the spectrum-sharing issues presently troubling the industry.

However, one constraint for GSO systems is the finite number of orbital slots, as shown in Figure 1.2.

The amount of orbital separation, set by the ITU, used to be 3° (120 orbital slots) and is now 2° (180 orbital slots) [21]. Any two GSO satellites are

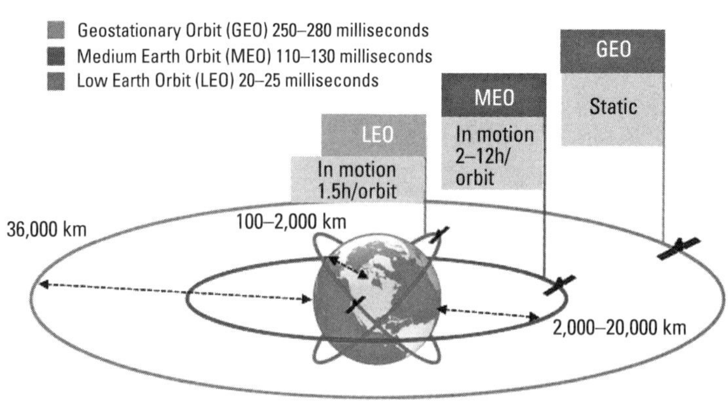

Figure 1.1 LEO, MEO, and GSO. (Image courtesy of Inmarsat and Euroconsult.)

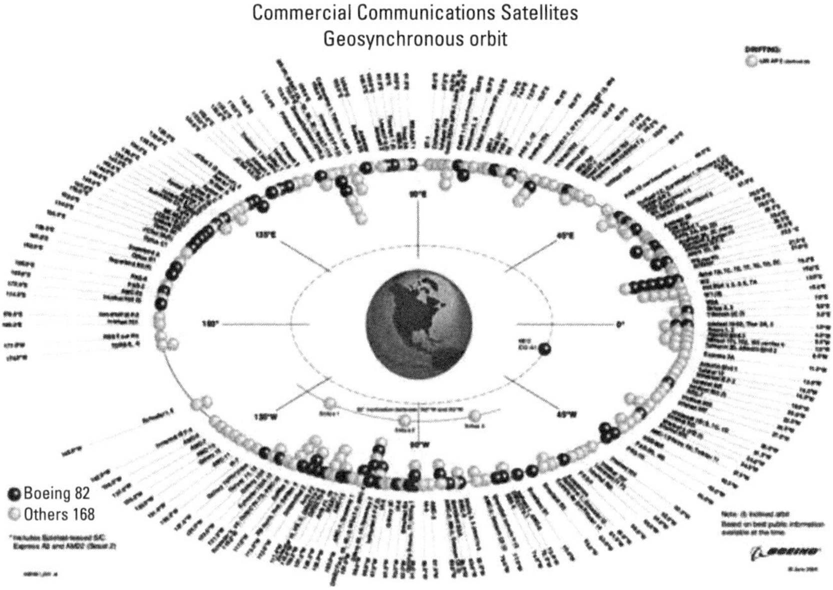

Figure 1.2 GSO slots [21]. (Thanks to the Boeing Corporation.)

separated from each other by about 75 km (45 miles), just slightly more than the diameter of Greater London [22].

Capacity can be increased by increasing the RF power and bandwidth of each satellite but this requires larger satellites. In the past, limits to rocket technology have made it hard (expensive) to increase the weight limit much beyond 6,000 kg (the largest GSO satellite is TerreStar-1 [6,910 kg], launched in 2016 on an Ariane 5 rocket); however, 10,000-kg payloads are now possible and the new generation of rockets being designed for deep space missions (to Mars and beyond) increase lift capability to more than 60,000 kg into low Earth orbit.

Colocated satellites (satellites that appear to be in the same place when viewed from Earth) increase GSO capacity and buddy SATs are now proposed in which additional satellites are sent to dock with existing satellites, doubling capacity and power for each unit addition. Work by the Defense Advanced Research Projects Agency (DARPA) developing the capability to perform through life repair, maintenance, and hardware upgrades of GSO satellites could also substantially improve GSO delivery economics [23].

Note that it is possible to provide east-to-west global coverage from four GSO satellites, although it is not uncommon for operators to own or lease transponder bandwidth on 40 or more satellites in order to deliver additional capacity, higher (and less variable) flux density. High GSO constellation counts (40 rather than four satellites, for example) ensure that a GSO satellite will be nearly always nearly overhead at the equator, maximizing the path link budget and minimizing path latency (having to point at a lower east/west elevation adds several thousand kilometers to a GSO path length).

MEOs (sometimes called intermediate circular orbits [ICOs]) are most commonly used for navigation, environmental monitoring, and some communications satellites. The orbital periods of MEO satellites range from about 2 to nearly 24 hours (Telstar 1, launched in 1962, orbited in MEO).

The most well-known and most widely used MEO constellations are the GNSS constellations. Very few of us drive anywhere without being connected to a satellite network. We take GPS, GLONASS, BeiDou, and (in the future) Galileo for granted but the GNSS MEO constellations at 20,000 km are all spectacular examples of contemporary space engineering.

The O3b system is an example of a MEO communications system: satellites orbit at a height of 8,000 km.

LEOs are used for higher-bandwidth communications satellites (taking advantage of the shorter path and hence lower signal path loss), and for environment-sensing and other scientific satellites that (using a polar orbit) repeatedly circle the Earth to build up detailed maps of particular parameters. A good example of this is Gravity Recovery And Climate Experiment (GRACE), which has been making detailed measurements of Earth's gravity field anomalies since its launch in March 2002. GRACE uses a microwave ranging system to accurately measure changes in the speed and distance between two identical spacecraft flying in a polar orbit about 220 km apart: small changes in gravitation are detected by minute changes in the distance between the two spacecraft.

Typical orbit heights for LEO communication systems are shown in Table 1.2.

LEO systems do not have any orbit slot constraints or indeed size and weight constraints. The International Space Station (ISS), for example, in orbit at 400 km is the size of a football field and weighs 408,000 kg, although it was built over a long period at significant expense. Note that the ISS communicates with Earth via the NASA (GSO) Near-Earth Network, so it is an early example (1998) of a mixed constellation LEO/GSO constellation with interconstellation switching.

Satellites have to obey the Newtonian Laws of Physics, so satellites closer to the ground will be traveling faster. More satellites are needed in LEOs to provide equivalent coverage to MEO and GSO satellites. For example, Iridium

Table 1.2
Orbit Altitude Comparisons

Orbcomm	775 km
Iridium	780 km
OneWeb	1,200 km
Globalstar	1,410 km

satellites travel at 17,000 miles per hour (27,000 km per hour) and have a horizon-to-horizon transit time of 8 minutes. For 70% of the time, there will be more than one satellite in view, although the satellite will only be directly overhead occasionally and for a short period of time.

GPS satellites travel at 8,700 miles per hour (14,000 km per hour). The higher speed of the Iridium satellites gives them a stronger Doppler signal. When combined with a higher flux density (signal strength) at ground level, this provides an alternative time and location system known as the Iridium Satellite Time and Location System but introduces a need for additional time alignment if alternative physical layers such as long term evolution (LTE) are used where all users have to arrive at the same time at the base station within the constraints of a time-domain guard band known as the cyclic prefix.

Satellites today include picosatellites weighing less than 1 kg, nanosatellites weighing less than 10 kg, microsatellites weighing between 10 and 500 kg, and macrosatellites (>500 kg) (see Table 1.3). CubeSats are nanosatellites that are constructed using a standard size and form factor with one unit being a $10 \times 10 \times 10$ cm cube, but with the potential for multiple units to be bolted together or potentially docked together in space.

At the other end of the size scale, Inmarsat I-5 Ka-band satellites are big macrosatellites with a launch mass of 6,100 kg, the body height of a double-decker bus, a solar array wing span of 33.8m generating 15 kW of power, and a xenon ion propulsion system for in orbit maneuvering.

The economics of delivering large and small satellites into space are being transformed by launch innovation, for example, reusable rockets from Space X, Europeanized Soyuz rockets and electric satellites (launched into interim orbits before floating up to their final orbit). Satellites are lasting longer and can potentially be refuelled and repaired in space.

As stated earlier, historically maximum available payloads on a single rocket have been of the order of 10,000 kg. The latest Falcon Heavy Rocket is capable of lifting 63,800 kg into LEO or 26,700 kg to a geosynchronous orbit suggesting that four I-5 satellites could be launched on a single rocket.

1.10 Satellite Technology Innovation: Fractional Beamwidth Antennas

The topic of technology innovation is a critical thread through this chapter and all subsequent chapters. One important innovation we will be looking at is fractional beam width antennas, antennas with a 3-dB beamwidth between 0.5° and 1.5° implemented typically as 12 to 100 spot beam arrays on a satellite. These antennas couple to a new generation of VSAT antennas on Earth-based fixed and mobile Earth-based devices.

Table 1.3
BIGSATS and SMALLSATS

Picosatellites (CubeSats?)	Nanosatellites	Microsatellites	Macrosatellites
<1 kg	<10 kg	<500 kg	≥500 kg

At this point, it is worth highlighting the difference between fractional beamwidth antennas and multiple-input multiple-output (MIMO) systems. Both approaches require highly linear transmit and receive paths to support phase shifting, both require adequate wavelength spacing between antennas but everything else is different.

MIMO systems are configured to produce multiple paths with each path separately modulated and channel coded (and amplified) to support high per user data rates with adequate multiplexing efficiency over short distances. Fractional beamwidth antennas are configured to deliver link budget gain from single narrow beam paths between a base station and user/IoT device.

MIMO systems exploit multipath. Fractional beamwidth antennas minimize multipath (and the associated delay spread). A well-designed fractional beamwidth antenna can produce more than 40 dBi of isotropic gain; the primary objective is to support moderately high data rates over long distances rather than superhigh data rates over short distances. Fractional beamwidth antennas are the single most important technology enabler for the present generation of high throughput satellites. They are used to focus on-demand RF energy on small geographic areas.

Fractional beamwidth antennas can also be used in terrestrial and satellite networks to focus on-demand RF energy on individual users or IoT devices. Used in conjunction with angular power separation techniques, covered in more detail in later chapters, these antennas are an important technology enabler for cost-economic, power-efficient wide area high data rate, high-mobility 5G terrestrial and satellite networks and potentially enable these networks to coshare the same spectrum.

1.11 FDD Dual-Use, Dual-Band Spectrum with Fractional Beamwidth Antennas

The other important difference between MIMO and fractional beamwidth antennas is that MIMO is more efficient when implemented in time division duplexed (TDD) spectrum as the uplink and downlink are reciprocal.

However, TDD systems do not deliver the same sensitivity as frequency division duplexed (FDD) systems and get less sensitive and less efficient with distance. In other words, TDD systems do not scale efficiently in wide area networks (WANs) and only work adequately well if all operators are cosited, which, given present competition policy, is largely impractical. The same applies in the satellite sector.

A typical Ka-band satellite FDD band plan at 28 GHz has four 250-MHz uplink channels between 28.35 GHz and 30 GHz paired with a downlink between 17.7 and 21.2 GHz. This is matched to a military band uplink at 30 to 31 GHz and a military downlink at 20.2 to 21.2 GHz.

The Ka-band payload of an Inmarsat Global Express satellite can be switched between military and commercial frequencies with the military bands supporting a range of high added-value applications including unmanned aerial vehicle (UAV) connectivity and control.

1.12 Present Launch Plans: Intelsat and Eutelsat

In 2009, Intelsat announced a $3.5 billion fleet investment and a hosted payload agreement with the Australian defense force followed in 2012 by plans for a new generation (known as the EPIC generation) of high throughput satellites. Two of these 6,500-kg satellites, built by Boeing, are capable of being launched from a single Ariane 5 rocket. The satellites have Ku-band transponders with services being targeted to aeronautical and maritime markets, treading on Inmarsat's traditional stamping ground.

Eutelsat has a 44-transponder Ku-band electric satellite (Eutelsat 7C) planned for launch in the third quarter of 2018 optimized to provide service to Sub-Saharan Africa and a Ka-band satellite built in Israel called AMOS (Affordable Module Optimized Satellite, but also a Jewish prophet) to be launched on a Space X rocket from Cape Canaveral.

Facebook has announced an agreement with Eutelsat to use this satellite to provide low-cost internet access to Africa using six of the AMOS Ka-band spot beams. The satellite GSO at 4° west will also provide coverage for the Middle East and Western, Central, and Eastern Europe, although some country-specific landing rights issues will need to be resolved.

1.13 People and Politics in the Satellite Industry

This brings us to politics and the people behind the politics and back to the 2016 American Space Renaissance Act proposed by Congressman Jim Bridenstein, a Congressman from Oklahoma, home of the Oklahoma Air and Space Port [24].

The Act envisions a renaissance of the military, civil, and commercial U.S. space industry. Citing Mr. Putin's investment in Glonass, Mr. Bridenstein makes the American case for military investment in "the ultimate military high ground," the need to invest in civil space missions including a Mars mission (27 NASA space missions have been canceled over the past 20 years at a cost of $20 billion) and a favorable regulatory environment for Mr. Musk at Space X and Mr. Branson at Virgin Galactic and their fellow travelers.

The bill is supported by EchoStar owned by Charles Ergan, who also owns Dish Networks and bought Light Squared stock in 2013 at a deep discount following the Light Squared Chapter 11 filing in May 2012. As with the early years of the competitively regulated cellular industry, individuals can make a major market impact and a not inconsiderable fortune, Craig McCaw being a notable example.

Mr. McCaw was a founding investor in the Teledesic satellite project, a planned constellation of LEO satellites operating in Ka-band (30-GHz uplink/20-GHz downlink) with the mission to deliver low-cost internet connectivity from initially 840 satellites (1993) and then 288 satellites (1997). Teledesic closed down in October 2002, having spent the best part of $1 billion. The spectral and orbital asset rights were acquired later by Greg Wyler for the O3b MEO network now owned and managed by SES.

By comparison, Light Squared (the company set up in 2010 to implement an L-band hybrid terrestrial satellite network) has reemerged from Chapter 11 as Ligado, the Spanish word for connected, chaired by Ivan Seidenberg, the former chairman and CEO of Verizon, and Reed Hundt, the former chairman of the FCC. The name at least suggests a Latin American low-cost interconnectivity business plan.

1.14 Third Time Lucky for Hybrid Satellite Terrestrial Networks?

The reappearance of Light Squared could be interpreted as a positive indication that hybrid terrestrial satellite networks could be on the agenda again. There are existing examples of hybrid networks such as Thuraya (GSM+ satellite) that are technically and commercially successful but only in high ARPU countries with large amounts of desert. There are also VHF satellite systems such as Orbcomm providing IoT connectivity that can be combined with terrestrial cellular networks. Orbcomm include a cellular modem in their service offer.

Dish Networks has applied for a patent for reusing frequencies between satellite and terrestrial systems based on MIMO and beam forming (see Section 1.18). Dish is part of a coalition of 10 companies that is lobbying the FCC to reallocate 500 MHz of presently unused Non Geostationary Orbit Fixed Service Spectrum (NGSO FSS) between 12.2 and 12.7 GHz (the lower end of Ku-band) for 5G Multi-Channel Video Distribution and Data Services (MVDSS). This would reduce the NGSO allocation to 11.7 to 12.2 GHz, although this is being contested by Space X, One Web LCC, and Intelsat.

Dish Networks also has access to cellular spectrum in the U.S. market. Inmarsat similarly has plans to implement a hybrid terrestrial and satellite network using their S-band spectrum adjacent to Band 1 implemented as a joint venture with Deutsche Telkom, and known as the European Aviation Network.

1.15 Scale and Standards Bandwidth

Hybrid satellite terrestrial networks, specifically hybrid satellite and 5G networks, will require significant additional work to be invested in the 5G nonterrestrial network standards process.

Many thousands of engineers spend many hundreds of thousands of man-hours producing 5G standards and specifications documents. However, the satellite industry is two orders of magnitude smaller and therefore has significantly less standards bandwidth available.

Apart from satellite TV with DVB-S as a relatively widely adopted standard in Europe and Asia, the satellite industry is dominated by proprietary physical layers with minimal overlap with present terrestrial cellular radio standards. Within ETSI, efforts were made 10 years ago to support UMTS/IMT-2000 interoperability with satellite systems at 2 GHz adjacent to terrestrial cellular Band 1, but little progress was made. Market presentations about LTE and satellite integration are statements of intention rather than imminent reality. However, we would argue that 5G needs the satellite industry and the satellite industry needs 5G and in both directions 5G/satellite integration will be mutually beneficial.

The satellite industry needs the 5G community because it needs access to consumer scale. 5G needs the satellite industry because economic wide area high data rate, high-mobility connectivity can only be achieved by using techniques such as adaptive fractional beamwidth antennas that are already deployed in the satellite sector combined with vertical coverage from nearly always nearly overhead satellites with line-of-sight visibility in to urban canyons and deep rural valleys and on to large open spaces. Economic wide area high data rate high mobility connectivity can only be achieved by using spectrum already used by the satellite industry and the satellite industry is arguably in the best position (literally directly overhead) to realize value from that spectrum. Direct line-of-sight links vertically up and down at elevation angles close to 90° are the only efficient way of avoiding the high surface absorption and ground reflections at higher frequencies. Conversely, high surface absorption and ground reflections in terrestrial networks with limited line-of-sight visibility will significantly compromise 5G delivery efficiency, a narrative that we revisit in later chapters.

1.16 Channel Bandwidths and Passbands: Satellite and 5G Band Plan Implications

Why do we need to use these higher frequencies? This becomes obvious when channel bandwidth requirements are considered.

As per-user data rates increase, wider channel bandwidths are required to maintain multiplexing gain. However, wider channel bandwidths reduce RF efficiency, particularly in space-constrained user devices.

For example, in an ideal world, antennas would work over a bandwidth of 10% of center frequency and filters would work over 4% of center frequency. These ideal bandwidths are often exceeded. With antennas, this is achieved by changing the electrical length of the antenna or increasing the physical length (for example, a planar inverted F-antenna (PIFA) [25]), but in both cases there will be an efficiency loss. With RF filters, wider passbands soften the filter edges and increase the amount of adjacent channel leakage and intersystem, and intrasystem interference. This can be mitigated by introducing additional filters, roofing filters as one example, but these increase insertion loss and take power out of the mobile uplink link budget.

The get-out clause is that it is not the channel bandwidth that is important but the channel bandwidth as a ratio of the center frequency.

RF filters are the reason that passbands in cellular FDD networks below 1 GHz are typically not more than 40 MHz (4% bandwidth ratio) supporting some combination of 5-MHz and 10-MHz LTE channels. The expectation within LTE Advanced is that passbands of 100 MHz will be needed to deliver an adequate compromise between multiplexing gain and RF efficiency. An efficient (3.3%) bandwidth ratio is 100 MHz at 3 GHz.

There is some consensus that an initial 5G network deployment in 2020 will need a channel bandwidth of 250 MHz to deliver adequate multiplexing efficiency. If spectrum continues to be auctioned on the basis of four operators per band this implies a passband of 1 GHz.

This means that the center frequency will need to be somewhere close to 30 GHz. This coincides with the spectrum presently being used by Ka-band HTS satellites. Anything much below this would compromise RF efficiency.

By 2025, a channel bandwidth of 500 MHz implies a passband of 2 GHz increasing to 1 GHz by 2030 implying a passband of 5 GHz. This bandwidth is only practical from an RF efficiency bandwidth ratio perspective using the millimeter band, with the spectrum either side of automotive radar being a potential option.

Automotive radar is being implemented between 77 and 81 GHz leaving 5-GHz passbands on either side between 72 and 77 GHz (immediately adjacent to the newly designated U.S. unlicensed band between 64 and 71 GHz) and 82 to 87 GHz. This assumes that in 10 years' time, digital signal processors will be capable of handling 1-GHz channel bandwidths and 5-GHz passbands power efficiently across a dynamic range of 100 dB. Given that the automotive industry has a similar problem to solve, it will probably happen.

Ku-band is a possible alternative to Ka-band and has the advantage of a lower fade margin, but it is currently hard to see how these proposals could scale globally. The passband of 500 MHz potentially available at 12 GHz is also arguably insufficient if a multi-operator auction model is required. A 1-GHz passband at 12 GHz, assuming it could be made available, would result in a loss of RF passband efficiency.

By contrast, the 28-GHz band is conveniently allocated on a 250-MHz channel raster within a 1-GHz passband with an efficient (2.5%) bandwidth ratio. The band has an established scale in fixed link terrestrial hardware, which could be translated into low-cost 5G hardware. Therefore, 28 GHz is arguably an optimum technical and commercial start point for 5G deployment with 38 to 40 GHz as a second alternative.

Later deployments based on 500-MHz and 1-GHz channel bandwidths within a passband of 4 or 5 GHz are going to be technically more efficient at millimeter wavelengths at 70 and 80 GHz.

It is difficult to see how 5G can be deployed cost-efficiently and power-efficiently without borrowing from present satellite technologies and without initially using satellite spectrum in the centimeter band (Ka-band and possibly Ku-band) and longer term in E-band, V-band, and W-band (the millimeter band).

AT&T announcements with EchoStar, Verizon, Viasat, Facebook, and Eutelsat are an early sign of this emerging dependency and appear to be validated by a shift in U.S. spectrum and competition policy.

This shift is not reflected in present ITU spectrum or standards policy and needs to be factored in to future competition policy. Satellite operators have been gifted their spectrum and typically have access to at least 4 GHz of aggregated bandwidth (including L-band and C-band allocations).

It will be a delicate balancing act to arbitrate what are likely to be complex coexistence, cosharing, cooperation, and commercial challenges and opportunities between the 5G community and satellite industry. The complexity will likely be compounded by the United States taking a significantly different approach to the rest of the world in terms of regulatory and competition policy. The 28-GHz band would appear to be particularly well suited to initial 5G deployment but will be politically challenging if global scale is to be achieved.

1.17 Impact of NEWLEOs Deployments: The Progressive Pitch Sales Pitch

From a regulatory perspective, satellites are divided into geostationary (GSO) and nongeostationary (NGSO). The ITU specifies that GSO satellites have priority over MEO and LEO (NGSO) satellites with regard to frequency usage. LEO satellites in (more or less circular) polar orbits between 160 and 2,000 km pass regularly between users and gateways on the ground and MEO and GSO satellites and therefore have to prove that they meet agreed coexistence criteria.

Over the past 20 years, Iridium and Globalstar have shown that it is eminently possible for LEO and MEO and GSO constellations to coexist but this is on the basis of narrowband (10 + 10 MHz) user links in L-band.

Iridium mitigate gateway to gateway interference in Ka-band by using intersatellite switching (between 23.187 GHz and 23.387 GHz). LeoSat are proposing a similar approach using the same Thales-based platform as Iridium. Some of the proposed NEWLEOs such as the Space X constellation propose to intersatellite switch using optical transceivers.

The substantive difference between Iridium and NEWLEO operators such as Space X, One Web, and LeoSat is the use of Ku-band for ground to space and space to user links.

OneWeb acquired the spectrum and access rights owned originally by Skybridge Incorporated, a United States entity established in the 1990s to roll out a high satellite count LEO constellation. The Ku-band passband for the downlink is between 10.7 and 12.7 GHz and the uplink is 12.75–14.5 GHz. The gateway downlink passband is 17.8–20.2 GHz with the downlink at 27.5–30 GHz. In the original FCC filing, Skybridge proposed to meet the U.S. Ku-band effective isotropic radiated power (EIRP) and flux density limits and protection ratios to the shared services supported in and adjacent to the passband by using progressive pitch angular power separation.

This means that as the satellites move towards the equator they deliver their power at a progressively more inclined angle to avoid sending power into GSO satellite receivers pointing directly upwards. As they move away from the equator, the power is delivered more directly downwards on the basis that GSO satellite dishes will be pointing at a progressively lower elevation.

This is achieved by slowly rolling the satellite in one direction then reversing the roll after passing the equator and switching off transmission when directly overhead. Given that the orbit time is 110 minutes, this happens every 55 minutes using reaction or momentum wheels powered from the solar panels on the satellite, a simple but clever system. We revisit this in more detail in Chapter 7.

The FCC was subjected to significant lobbying from other incumbent users in the Ku user and Ka gateway passbands with the methodology used to calculate interference levels cited as a major concern.

Twenty years on, these arguments continue. OneWeb, Space X, and LeoSat stress that their progressive pitch approach, coupled to adaptive power control and in some case fractional beamwidth adaptive antennas, is significantly more effective than the original Skybridge (and Teledesic) proposals, but the modeling is significantly complex, particularly when multiple constellations sharing the same passbands have to be taken into account. There are also a wide range of potential victim receivers ranging from high definition and ultrahigh definition (UHD) satellite TV, very small aperture terminals and a wide mix of civilian and military two-way radio systems.

Conversely, if relatively extreme inclination angles are imposed on the NEWLEOs, there will be a directly adverse impact on the link budget, additional latency, and a capacity cost, all of which will subtract value from the NEWLEO business model.

There is another potentially tricky aspect to the progressive pitch sales pitch. If the NEWLEOs can demonstrate that they can coexist with GSO operators in the same passbands, then it could also be assumed that the spatial separation and power techniques used to achieve this could be equally effective in allowing 5G operators to coshare the spectrum, including, for example, the 28-GHz band.

This could form the basis of some interesting technical and regulatory arguments at WRC 2019 and brings us back to the topic of regulatory and competition policy.

1.18 Flat VSATs: An Alternative to Progressive Pitch as a Mechanism for Cosharing 5G and Satellite Spectrum

In Chapter 6, we explore an alternative approach to managing coexistence and in band-sharing based on low-cost Flat VSATs and passive and active flat panel and conformal arrays including high-element count arrays (256/512/1,024 elements) integrated into TV displays for indoor coverage and outdoor coverage via advertising hoardings and the same arrays integrated into solar panels for high throughput and very high throughput terrestrial outdoor mobile and fixed access.

We also explore the potential for 16-element and 32-element arrays embedded into the screen of smartphones and wearable devices and show how this could deliver a share of smartphone-connected added value to the satellite industry.

This seem to us to be a safer and more robust approach to spectrum sharing and provides the admittedly contentious prospect of cosharing 12 GHz, 28 GHz, and V-band and E-band passbands between 5G, 5G in band backhaul, and LEO, MEO, and GSO constellations. It also opens up the opportunity to reuse existing cellular spectrum from 450 MHz to 3.8 GHz for satellite connectivity.

1.19 Coexistence and Competition, Subsidies, and Universal Service Obligations

We have said that spectrum access rights and, in the satellite industry, orbit access rights and landing rights are conferred on the basis of expected and promised social and economic benefits from improved connectivity. This could either be benefits delivered to consumers or corporate and industrial users, public safety and disaster relief, and emergency services or to military and defense communities. The promised benefits are predicated on various combinations of technology and commercial innovation or improved exploitation of the underlying properties of a delivery medium.

For example, as referenced earlier, LeoSat has a distinctive business model based on the proposition that radio waves move faster in free space than light in fiber. By contrast, OneWeb and Space X in their FCC filings stress their potential role in connecting the unconnected or underconnected. Depending on how you count them, this amounts to about 35 million people in the United States and 3 to 4 billion people globally.

Greg Wyler, the founder of O3b (the Other Three Billion), successfully used this argument to gain regulatory approval for the O3b MEO constellation in 2008, having acquired Ka-band spectrum from Teledesic when it stopped constellation development in 2002. This provided O3b with access rights to the downlink passband between 17.7 and 20.2 GHz and an uplink between 27.5 and 30 GHz, the same bands that are proposed to be used for the NEWLEOs Ka-band gateway uplinks and downlinks and already used by Iridium and a number of GSO operators.

O3b inconveniently had to raise capital in the year that Lehmann Brothers went bankrupt, and it is a tribute to the persuasive skills of the Wyler management team that the constellation launched and more or less met its business plan objectives. However, it achieved this by substantially changing the market focus of the business, which now supplies internet connectivity to cruise ships 40° either side of the equator (the industry joke was that O3b stood for Only Three Boats rather than the Other Three Billion).

This highlights the problem that many of the presently disconnected are low-income or no-income customers so making any comprehensive inroads into the digital divide is likely to require substantial government subsidy on a country-by-country, region-by-region basis.

This already happens with terrestrial fiber subsidies or via universal service obligations imposed with various financial incentives. The amount of digital divide subsidy going to the satellite industry is relatively small (of the order of 1.5% in the U.S. market) and the NEWLEO contenders including Space X make a persuasive argument that these dollars would be more effectively spent with them rather than on terrestrial system subsidies.

Whether this is the case depends on the fine detail of the final agreements on coexistence with the agreement process now made more complex by the ambitions of the 5G community to share or acquire Ku-band and Ka-band spectral assets.

This includes a growing recognition by aspiring 5G operators that the principle of angular power separation could be applied to support cosharing between terrestrial 5G and LEO, MEO, and GSO networks, a combination that would provide superlative global coverage and capacity gain achieved through spatial frequency reuse.

However, there are substantial regulatory barriers that need to be overcome before this becomes a practical proposition. The failure of the proposed Intelsat and OneWeb merger provides a case in point. It may have been that the Intelsat bond holders were wary about increasing their gearing ratios, already stratospherically high. It may also have been influenced by a nagging worry that Intelsat's spectrum and international landing rights, patiently negotiated over 50 years, could have been open to legal challenge if the merger had gone ahead.

It may be that this particular logjam will be unlocked by Google or Facebook. As stated earlier, Google and Fidelity Investments already have a 10% stake in Space X in return for a $1 billion investment and both companies have enough spare cash to buy a large part of the satellite industry at present enterprise value. They also have substantial regulatory influence.

In the meantime, it is important that the NEWLEOs do not shoot themselves in the foot by disputing each other's interference models and offer a unified vision to regulatory authorities around the world based on the thesis that high-count LEO satellite constellations have a critical and economically compelling role to play in future internet connectivity and can and should be seamlessly integrated with MEO, GSO, and terrestrial 5G networks.

1.20 U.S. Competition and Spectral Policy

In April 2016, AT&T and EchoStar announced a potential sharing framework for the 28-GHz band with Hughes Network Systems and Alta Wireless as possible partners. In parallel, Verizon and Viasat agreed to undertake coexistence, cosharing, and cooperation studies. This followed the AT&T filing with the FCC in January 2016 for an experimental license to conduct fixed and mobile testing with various types of new wireless equipment between 27.5 GHz and 28.5 GHz.

The CTIA suggested these studies should be broadened to include Upper Microwave Flexible Use (UMFU) shared access agreements in the 37 to 40 GHz band. In July 2016, the FCC responded by approving its Spectrum Frontiers proceeding releasing UMFU designations for 27–28.35 GHz, 37–38.6 GHz, and 38.6–40 GHz and a new unlicensed band at 64–71 GHz.

The interest by the United States and other potential 5G terrestrial mobile operators in the 28-GHz and 38–40 GHz bands is easy to explain. The satellite industry has access to FDD spectrum, which is ideally suited to terrestrial 5G implementation. Additionally, the satellite industry has successfully implemented fractional beam width antenna technology, which meets many and potentially all of the 5G wide area high data rate link budget requirements.

Additional scale economy benefits are also realizable from 28-GHz and 38-GHz terrestrial fixed link hardware. However, many satellite operators including Southeast Asian operators are opposed to the idea of having 5G in the 28-GHz band and are suggesting that investment in high throughput satellites using or proposing to use this band will be compromised. It is not unusual for the United States to have a different regulatory outlook, and the future of the 28-GHz band will almost certainly continue to be debated vigorously for the foreseeable future.

1.21 Satellites and Local Area Connectivity

NEWLEO business models are based on the assumption that Wi-Fi will be used to provide local connectivity from small, optionally solar-powered, low-cost base stations and there may be similar opportunities to integrate Bluetooth into these optimized localized delivery systems.

We cover Wi-Fi 802.11, 802.15, proprietary low-power drain, long-distance licensed radio options such as SigFox and Lora and latest iterations of Bluetooth including low-energy Bluetooth and long-distance Bluetooth in Chapter 10, but on a general note, it can be observed that the integration of local area and personal area networks with 5G and satellite systems will be one of the critical paths to delivering a satisfactory 5G user and IoT experience. Part of the challenge will be managing integration of these enhanced legacy radio systems into low-cost, small-form factor user and IoT devices.

Traditionally, physical layer design for 3G and 4G networks has been focused on delivering in-band spectral efficiency using higher-order modulation with substantial envelope modulation. These options trade off in-band spectral efficiency against spectral splash into adjacent out-of-band spectrum. They are also not inherently power-efficient.

5G includes waveforms that are sufficiently power efficient and sufficiently narrowband for battery-powered IoT applications. These need to be regarded as complementary rather than competitive to other optimized legacy technology options.

In the context of 5G and satellite integration and in an ideal world, 5G and satellite would use the same physical layer or at least share some baseline commonality. This remains possible as the Release 16 and 17 standards process has a measure of flexibility in terms of physical layer implementation for Ku-band, K-band, and Ka-band, although as we highlight in Chapter 10, issues of standards integration are only just now being addressed with significant work still to be done before any meaningful integration is achieved.

It may be more likely that some compatibility can be achieved between the Low Mobility Large Cell (LMLC) 5G physical layer and satellite physical layer particularly as satellites have the potentially useful capability to scale from the typical maximum cell size achievable in terrestrial mobile broadband systems (between 35 and 100 km) up to cell sizes of 2,000 km or more.

1.22 Summary

Over the past 60 years, there has been a steady consolidation of spectral access rights in the satellite industry scaling from VHF through L-band, S-band, and C-band to Ku-band, K-band, and Ka-band with submissions now being made for V-band and W-band allocations. These spectral access rights are coupled to intensively negotiated orbital rights and country by country access rights.

NEWLEO operators either have to undergo these regulatory processes in a much reduced time scale (5 years rather than 50 years) or merge with established GSO, MEO, and LEO operators with existing rights. However, such mergers might be subject to legal challenge and in practice regulatory and competition policy barriers might be more problematic than the technical challenges ahead.

The NEWLEO operators are confident that they can reduce delivery costs by at least an order of magnitude compared to existing operators and can cost-effectively provision sufficient capacity to meet the demand that these lower-cost points could realize. A fast rate of rate decline would be problematic for some existing GSO operators that are reliant on generous margins to meet present debt cover commitments.

Cash-rich over-the-top (OTT) players such as the Google, Apple, Facebook, and Amazon (GAFA) quartet could help resolve this commercial tension.

An initial investment by The Alphabet Group of £1 billion in Space X suggests a developing interest from Web-scale majors in the satellite sector.

There are some things that can only be done from a satellite, some things that are better done from a satellite, and some things that are better done locally within a terrestrial network.

Superficially, it seems daft to send a signal hundreds or thousands of kilometers into the sky when compared to the option of a base station a few meters away, but network densification in 4G and 5G is increasing routing complexity and the cost and power drain of backhaul networks is making the end-to-end journey unpredictable and occasionally expensive.

There is therefore an increasingly persuasive argument that an increasing amount of direct user and device traffic and indirect backhaul traffic in 4G and 5G networks could be carried more cost-effectively over satellite networks.

However, achieving global coverage requires a mix of LEO, MEO, and GSO constellations. Angular power separation potentially allows frequency reuse across these multiple constellations and reuse for 5G terrestrial point-to-point and user-to-base station links, but many issues of regulatory and competition policy need to be resolved before this becomes a practical reality.

In particular, the satellite industry and satellite industries remain locked into an adversarial spectral allocation and auction process that inhibits and frustrates cooperation between the two operator sectors and their respective supply chains.

This brings us to the subject of our next chapter, the race for space spectrum and the potential battleground issues that need to be resolved at WRC 2019 and WRC 2023.

References

[1] https://www.space.com/37200-read-elon-musk-spacex-mars-colony-plan.html.

[2] https://www.space.com/38314-elon-musk-spacex-mars-rocket-earth-travel.html?utm_source=notification.

[3] https://www.blueorigin.com/.

[4] http://www.virgingalactic.com/.

[5] https://www.facebook.com/Internetdotorg/.

[6] http://www.spacex.com/news/2015/01/20/financing-round.

[7] http://www.planetaryresources.com/#home-intro.

[8] http://www.goldmansachs.com/our-thinking/podcasts/episodes/05-22-2017-noah-poponak.html.

[9] http://www.hellenicaworld.com/Greece/Technology/en/CatapultTypes.html.

[10] https://www.nasaspaceflight.com/2017/09/spacex-first-x-37b-launch-falcon-9/.

[11] http://spacerenaissanceact.com/satellite-industry/.

[12] http://www.sciencemag.org/news/2017/06/china-s-quantum-satellite-achieves-spooky-action-record-distance.

[13] http://www.techdigest.tv/2017/05/middlesbrough-tops-table-4g-table-that-is-but-4g-coverage-remains-well-short-of-ofcom-target.html.

[14] http://www.bournemouthweather.co.uk/climate.php.

[15] https://www.ses.com/networks.

[16] http://oneweb.world/board/greg-wyler.

[17] https://www.ispreview.co.uk/index.php/2016/11/uk-government-pledge-1bn-boost-fuel-pure-fibre-optic-broadband.html.

[18] http://www.satellitetoday.com/newspace/2017/06/23/leosat-joins-forces-esa-satellite-5g/.

[19] http://www.sciencedirect.com/science/article/pii/S0273117714005766.

[20] http://qzss.go.jp/en/.

[21] http://www.satellitetoday.com/publications/via-satellite-magazine/features/2008/03/01/hot-orbital-slots-is-there-anything-left/

[22] https://space.stackexchange.com/questions/2515/how-closely-spaced-are-satellites-at-geo.

[23] https://www.darpa.mil/program/robotic-servicing-of-geosynchronous-satellites.

[24] https://airspaceportok.com/.

[25] http://www.antenna-theory.com/antennas/patches/pifa.php.

2

The Race for Space Spectrum

2.1 Why Spectrum Is Important

Our last book, *5G Spectrum and Standards*, was completed just after the end of WRC 2015. If you have just bought a recent copy of this current book, then it is probably just before the next World Radio Conference to be held in 2019. If you are reading this after the WRC 2019 Conference, then at least you can form a view of whether things turned out as we said they would or should.

The spectrum allocation and auction process is essentially adversarial and designed with the theoretic objective of maximizing social and economic and occasionally political gain from spectrum as a finite although reusable and shareable asset. There are dozens of stakeholders involved, for example, the World Meteorological Organization is worried about a number of issues [1].

World Radio Conferences are huge events requiring hundreds of thousands of hours of preparation time [2]. Figure 2.1 shows delegates at the opening session of the 2015 event.

Spectrum is important for satellite systems because the quantity and quality of the spectrum made available and the usage conditions and or service and or coexistence obligations imposed on the users of that spectrum determine the capacity and coverage and hence economics of the system.

Coexistence includes the management of interference between GEO, MEO, and GSO satellite systems cosharing the same bands or adjacent to

Figure 2.1 2015 World Radio Conference opening session, Geneva. (© ITU/D. Woldu.)

other bands and interference between two-way communication systems and Earth-based satellite TV receivers.

2.2 5G Coexistence with Satellite TV and Other Satellite Systems

Satellite TV is deployed into C-band between 3.7 and 4.2 GHz, into Ku-band between 8 GHz and 12 GHz (predominantly between 11.7 and 12.7 GHz, which is what most of us watch at home), and in Ka-band at 18.3–18.8 GHz and 19.7–20.2 GHz. The Ka-band allocation (18.3–18.8 GHz + 19.7–20.2 GHz) is used for superhigh-definition and ultra-high definition TV.

C-band typically supports 250 channels of video and 75 audio services using dishes, which average 2m in diameter. C-band dishes are steerable, enabling C-band users to receive signals from 20 or more satellites. Fifth generation (5G) operators are keen to deploy 5G terrestrial services into this spectrum. The 5G community often states that C-band TV is rapidly disappearing, but it lives on in a surprising number of countries. My neighbor has a large C-band dish pointing at the horizon that I assume he uses to watch Turkish TV and it would be perfectly possible for a 5G C-band signal from a mobile or base station to pour unwanted energy into his satellite TV front end. A 5G C-band network in Singapore between 3.7 and 4.2 GHz would need to coexist with C-band TV receivers in Malaysia, although at least the TV dishes will be pointing more or less directly upwards.

The Multichannel Video Distribution Service (MVDDS) coalition [3] in the United States is currently lobbying the FCC to reexamine technical limits in the 12.2–12.7-GHz band so that it can offer two-way mobile broadband services instead of one-way fixed service as currently permitted. The U.S. company

Dish Networks is an active member of this advocacy group. Dish Networks supply satellite TV in all three TV bands (C-band, Ku-band, and Ka-band) and have access rights to three terrestrial cellular bands: unpaired AWS-3 uplink spectrum (1,695 MHz to 1,710 MHz), H Block downlink spectrum (1,995 MHz to 2,000 MHz), and AWS-4 spectrum (2,000 MHz to 2,020 MHz).

In June 2016, 3GPP formally approved Band 70, which aggregates these terrestrial bands together. This opens up the possibility that Dish Networks could realize a tri-band LTE terrestrial network integrated with satellite TV and two-way satellite services in Ku-band, although this will require regulatory approval. These bands are not universally available in other markets and it would be unlikely that other mobile broadband terrestrial operators would have the same or similar terrestrial aggregated band plan suggesting that the proposal might be constrained by a lack of global scale. Sprint is another U.S. operator-specific example. The Sprint Gigabit LTE tri-band proposal combines their 800-MHz, 1,900-MHz and 2.5-GHz band allocations. Sprint underwent a major recapitalization in 2012 largely financed by Softbank [4]. Softbank is also a major investor in OneWeb suggesting that cross-investment in satellite and terrestrial properties including mobile broadband and traditional broadcasting might become more common place.

Being financed by the same bank does not however solve coexistence issues. Many GSO satellite operators support a mix of TV transponders and two-way communication services so they can actively manage any in-band or adjacent band interference. Coexistence with Ku-band LEO downlinks is more problematic and needs to be managed through angular power separation and polarization diversity (both covered in more detail in Chapters 5, 6, and 7).

The opportunity for disputes between LEO, MEO, and GSO satellite operators are therefore many and various. Adding 5G territorial operators to the mix makes an already complicated picture more complex.

2.3 Radar Frequency Band Designations

Frequency bands for the satellite industry (and for fixed point-to-point backhaul and for 5G terrestrial) are described using the IEEE Standard 521-1984 Radar Frequency Band designations, as shown in Table 2.1.

Satellites can be found right through the electromagnetic spectrum from very high frequency (VHF) through to V-band and W-band (and higher for some military communication systems). For example, the Orbcomm constellation [5] provides narrowband IoT connectivity in the VHF band, Iridium [6] and Globalstar [7] are implementing their second-generation LEO constellations in L-band and S-band and the NEWLEOs coshare spectrum for user

Table 2.1
IEEE Standard 521-1984 Radar Frequency Bands

L-Band	S-Band	C-Band	X-Band	Ku-Band	K-Band	Ka-Band	V-Band*	W-Band*
1–2 GHz	2–4 GHz	4–8 GHz	8–12 GHz	12–18 GHz	18–27 GHz	27–40 GHz	40–75 GHz	75–110 GHz
GPS	MSS	TV	Military	Commercial	Military	Commercial	Military, Commercial, and Automotive Radar	
Licensed	Licensed	Licensed	Licensed	Licensed	Licensed	Licensed	Unlicensed	

*The description E-band is also sometimes used to describe a large subband between 60 and 90 GHz. You may also come across Q-band as a designation which like E-band comes from the WR22 waveguide naming system. Q-band covers from 33 GHz to 50 GHz (9.1-mm to 6-mm wavelength).

uplinks and downlinks in Ku-band, gateway uplinks and downlinks at K-band and Ka-band and telemetry and telecontrol links in K-band.

2.4 5G Standards and Spectrum

The 3GPP Release 15 standards process defines possible band plans for terrestrial 5G below 6 GHz (in C-band, for example). The 3GPP Release 16 standards process defines possible band plans for terrestrial 5G in Ka-band and E-band. HTS Ka-band satellites (high-capacity GSO satellites) are normally deployed as a frequency division duplex (FDD) with 250-MHz channel spacing in typically a 3.5-GHz passband. This is ideal for 5G.

The satellite industry is unhappy at the prospect of losing primary access rights to Ku-band and Ka-band spectrum and at WRC 2015 successfully limited the options for study for WRC 2019. These are shown in Table 2.2. This includes FCC proposals to consider Upper Microwave Flexible Use as a mechanism for cosharing Ku-band, K-band, and Ka-band spectrum [8]. Note that the satellite industry is particularly unhappy at the prospect of having 5G in the 28-GHz band. This is simply explained by the list of present high throughput satellites using this band, which includes 28-GHz band high throughput satellite GSO incumbents and Australian National Broadband Network satellites (2 satellites), IPStar (4 satellites), Inmarsat Global Xpress (4 satellites), O3b MEO (12 satellites), Viasat (4 satellites), Jupiter (2 satellites), Hylas/Avanti (2 satellites), Amazonas 3, Spaceway 3, Wild Blue1, Superbird4, AMC 15 and 16, and a number of direct TV satellites.

There is also a proposal to use E-band either side of the 77-GHz automotive radar band. The probable band plan is shown in Table 2.3.

Further study is needed to quantify potential interference issues within the automotive radar band and the adjacent passbands. Multiple radar systems operating between 76 and 81 GHz imply significant spectral density and strong pulsed signals, which could potentially cause in-band and out-of-band interference.

The lower band edge of the lower duplex (71–76 GHz, E-band) is also immediately proximate to the proposed extended 60-GHz Wi-Fi band, which will potentially yield around 15 GHz of contiguous unlicensed spectrum.

From a satellite perspective, the significance of the extended 60-GHz band is that, together with the 2.4-GHz and 5-GHz Wi-Fi band, they produce a no-cost or more accurately low-cost connectivity solution, which can be integrated with NEWLEO system solutions. OneWeb provides one example of this proposed approach. There is also a proposed extension of the 5-GHz Wi-Fi band for automotive connectivity.

Table 2.2
Bands Agreed for Study at WRC 2019 (Not Including the 28-GHz Band)

ITU WARC 2109 Bands Agree for Study for 5G													
	K-Band		**Ka-Band**			**V-Band**						**W-Band**	**Total**
GHz	24.25 27.5		31.8 33.8	37 40.5	40.5 42.5	42.5 43.5	45.5 47	47 47.2	47.2 50.6	50.4 52.6	66 76	81 86	
GHz	3.25		1.6	3.5	2	1	1.5	200 MHz	3	2.2	10	5	33.8 GHz
FCC Upper Microwave Flexible Use													
GHz			27.5 28.35	37 38.6	38.6 40						64 71		
GHz			850 MHz	1.6	1.4						7 GHz Unlicensed		
			Licensed	Licensed*									10.85 GHz
GHz				FCC ITU									
Summary	33.8 GHz of ITU spectrum for study at WRC2019 10.85 GHz of FCC UMFU spectrum for study of which 3 HGz is common (37-40 GHz) FCC proposed lower band at 28 GHz not included as an ITU WRC 2019 stufy band through adjacent to ITU study band *600 MHz of FCC spectrum from 37 to 37.6 GHz proposed as shared use commerical/federal												
FCC Future Notice of Proposed Rule Making													
GHz	24.25 24.45	25.5 25.25	31.8 33.4			42 42.5			47.2 50.2	50.4 52.6	71 76	81 86	
GHz	200 MHz	200 MHz	200 MHz			500 MHz			3 GHz	2.2 GHz	5 GHz	5 GHz	17.7 GHz
	FCC/ITU					FCC/ITU			FCC/ITU	FCC/ITU			
GHz													
Summary	17.7 GHz of FCC spectrum for study of which 17.7 GHz is common to ITU and FCC (24 GHz, 25 GHz, 32 GHz, 42, 47-50, 50,52 GHz, 71-76, 81-86 GHz IEEE 521-1985 radar bands—X Band 12–18 GHz, Ku Band 12–18 GHz, K-Band 18-27, Ka-ban 27-40 GHz, V band 40-75 GHz, W-band 75-110 GHz												

Table 2.3
5G E-Band

5G PPP E band channelization and coexistance											
CEPT											
71-76 GHz			76-77 GHz	77-81 GHz	81-86 GHz			86-92 GHz	92-95 GHz		
5G MOB TX?					5G MOB RX?				5G TDD		
Guard band	Channels	Guard band	Narrowband long range radar	Wideband short range radar	Guard Band	Channels	Guard band	Radio astronomy band	Guard band		Guard band
125 MHz	19 X 250 MHz	125 MHz			125 MHz	19 X 250 MHz	125 MHz		125 MHz	11 X 250 MHz	125 MHz
U.S. FCC					U.S. FCC						
4X 1.25 GHz channels					4 X 1.25 GHz channels						
Legacy use											
71-74 GHz	74-76				81-84 GHz	84-86 GHz					
Fixed satellite (space to Earth Mobile satellite (space to Earth)	As 71-74 plus Broadcasting Satellite Space Research (space to Earth)				Fixed satellite (Earth to space) Mobile satellite (Earth to space) Space research (Earth to space) Radio astronomy						

It has always been a challenge keeping up with the alphabetic progress of 802.11, but the latest 802.11ax chip sets [9] claim to support a headline data rate of 4.8 Gbps compared to the 1.7 Gbps available from an 802.11ac access point with multiple radios theoretically capable of supporting 10 Gbps of throughput and or up to 400 users per cell (fairly obviously not 400 users at 10 Gbps).

It is an apparently small detail, but the work on 5G frame structures within 3GPP working group RAN 1 includes the specification of mini slots consisting of a minimum of two symbols within a 1-ms time frame. This is partly to support Ultra Reliable Low Latency Communication (URLLC) and URLLC pre-emption in an eMBB (enhanced mobile broadband) channel but is also intended for operation in unlicensed bands, for example, to start transmission directly after a successful listen before talk procedure without waiting for a slot boundary.

The integration of Wi-Fi with LTE Assisted Access (LAA) and LTE-U and 5G should ideally also take into account potential satellite use of these bands, a topic that we revisit in Chapter 10.

2.5 Existing LEO L-Band, Ku-Band, K-Band, and Ka-Band Allocations

The existing allocations for Iridium and Globalstar at L-band are adjacent to Inmarsat L-band spectrum, the Iridium feeder gateway downlink is in K-band at 19.4–19.6 GHz with the gateway feeder uplink in Ka-band at 29.1–29.3 GHz and the intersatellite switching allocation at 23.187–23.387 GHz in K-band.

It is possible that the Legacy LEOS that use intersatellite switching, for example, Iridium, could obtain permission from the FCC and ITU to use their intersatellite and Earth station uplink and downlink spectrum in K-band and Ka-band for general wide area coverage using angular power separation to support frequency reuse and coexistence. This would transform Iridium's service offer. However, they would also need to scale their constellation to hundreds or thousands of satellites in order to have sufficient RF power and "nearly always nearly overhead" visibility to support mass market consumer and or low average revenue per user (ARPU) mobile and fixed access internet connectivity. Iridium has not made any announcements about this and their existing constellation upgrade is probably too advanced to be able to support a change in business model. They would seem to be well positioned to continue to service their traditional high-value subscribers effectively and efficiently. Note that intersatellite switching reduces the number of Earth gateways needed, reduces latency, and is arguably more power and bandwidth efficient and as stated earlier allows Iridium to support high added-value military payloads alongside their commercial offering.

Iridium gateway links are within the passband of the proposed OneWeb and LeoSat user links and O3b/SES Ka-band MEO downlink. As existing incumbents, and probably equally important as an operator carrying critical military payloads, it would be unlikely that the FCC would wish to impose any coexistence requirements on Iridium and far more likely that new market entrants will be required to meet stringent protection ratios to ensure existing and next generation Iridium service levels can be maintained.

In an ideal world, the mobile broadband community and satellite industry would work together to integrate band plans and technical standards and achieve mutual scale benefits. In practice, mobile operators, particularly U.S. mobile operators, are lobbying for primary access to existing satellite radio bands including spectrum in Ku-band, Ka-band, and E-band, an adversarial process that discourages cooperation.

As stated, there are existing fixed and mobile systems in L-band and S-band including LEO systems (Iridium and Globalstar) and GSO satellites (Inmarsat 4, for example). Satellites are also intensively deployed into licensed spectrum at C-band (4–8 GHz) including satellite TV (also at 10 GHz), into X-band (8–12 GHz), Ku-band (12–18 GHz), K-band (18–27 GHz), and Ka-band (27–40 GHz). The spectrum is coshared with military satellite systems, although many of these are presently concentrated in X-band and K-band.

A satellite operator can typically accrue several gigahertz of spectrum across these bands. A mobile operator, by comparison, will have at most 200 or 300 MHz across the UHF band and L-band, S-band, and lower end of C-band (TDD bands 42 and 43).

Table 2.4 shows how these bands fit in to the larger spectrum picture described in terms of wavelength.

The bands of particular interest are the meter band (from 300 MHz to 3 GHz), the centimeter band (3 to 30 GHz), and the millimeter band (30 to 300 GHz) also known as the Sub 10 band (wavelengths of 10 mm or below). See *5G Spectrum and Standards* for a more detailed analysis of this.

2.6 Benefits of Higher Frequencies/Shorter Wavelengths

The significance of the shorter wavelength bands is that it is possible to construct compact phased array antennas that deliver isotropic gain, offsetting the propagation loss at these higher frequencies.

These antennas are becoming widely used in the satellite industry particularly in Ku-band, Ka-band, and E-band. Antennas known as fractional

Table 2.4
Frequency and Wavelength Comparisons

	kHz	**kHz**	**kHz**	**MHz**	**MHz**	**MHz**	**GHz**	**GHz**
Frequency	3–30	30–300	30–3,000	3–30	30–300	300–3 GHz	3–30	30–300
Wavelength	Kilometers	Kilometers	Meters	Meters	Meters	Meters	Centimeters	Millimeters
	100–10	10–1	1,000–100	100–10	10–1	1–0.1	10–1	10–1
Name	100-km band	10-km band	Kilometer Band	100-m band	10-m band	Meter Band	Centimeter Band	Millimeter Band
		Long Wave	Medium Wave	Short Wave		Microwave		
Atmospheric noise up to 20 MHz								
			Galactic noise up to 100 MHz					
					Circuit noise			
					Aperture gain offsets propagation loss			

beamwidth antenna arrays with a beamwidth of between 0.5% and 1.5% can deliver a gain of more than 40 dBi. The additional propagation loss at 28 GHz compared to 900 MHz (low-band cellular) is of the order of 30 dB. The antennas can track moving satellites (LEO and MEO) minimizing pointing loss.

Satellites share their spectrum with deep space communication and commercial and military and weather radar at 2.7–2.9 GHz, 5.2–5.7 GHz, micro rain radar at 24 GHz in the water vapor resonance peak (shared with automotive radar), and cloud composition radar (cloud radar) at 35 GHz. Military applications include the telemetry and telecommand of unmanned aerial vehicles, high-definition imaging and surveillance, and remote weapon systems including anti-missile systems.

Each new generation of military and civilian satellite radio and radar system requires more rather than less bandwidth, increased transmit power and increase receive sensitivity. These requirements translate into the need for higher protection ratios (i.e., the ability to reject out-of-band signals) [10].

2.7 Spectrum: Why Ka-Band Is Useful

Figure 2.2 summarizes why Ka-band is a preferred band in terms of spectrum availability (3.5 GHz of presently available spectrum). It also highlights the potential of Q-band and V-band (and W-band/E-band not included in the graphic).

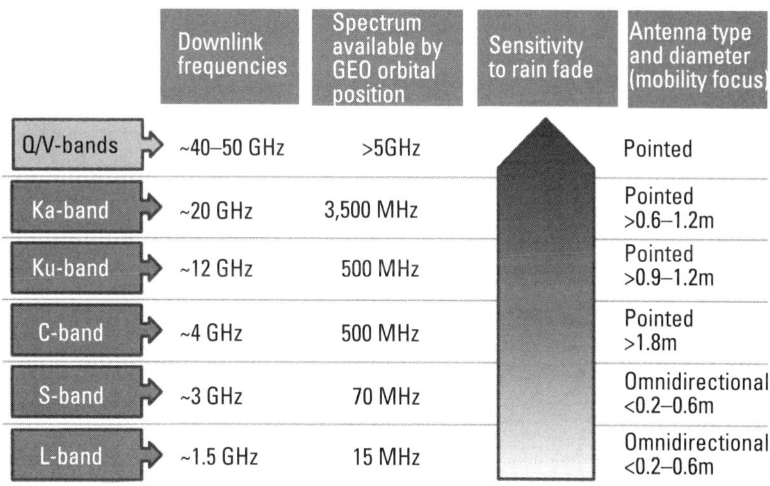

Figure 2.2 Five Ka-band and other band comparisons. (Thanks to Euroconsult and Inmarsat.)

2.8 The Impact of Standards on 5G Spectrum Requirements

The standards process also has an impact on spectral policy and potentially spectral demand. 3GPP Phase 1 Release 15, due in late 2018 concentrates on sub-6 GHz (including 3.8 to 4.99 GHz). Release 16 (Phase 2) includes 28-GHz and 38-GHz beam forming and was due to be completed in December 2019, by which time the outcome of the 2019 World Radio Conference will be known (Table 2.5). The first iteration of what is called 5G Non-Standalone New Radio was completed in December 2017.

Figure 2.3 shows the present 3GPP enhanced mobile broadband (eMBB) standards time line.

2.9 Multiplexing, Modulation, and Coexistence

The 5G New Radio layer uses what is called a flexible numerology. What this means is that different orthogonal frequency division multiplexing (OFDM) subcarriers can be chosen depending on the required application starting with 15 kHz, then 60 kHz, then 120 kHz, then 240 kHz, and 480 kHz. Figure 2.4 suggests 15-kHz subcarriers for use in FDD and TDD spectrum below 3 GHz for large outdoor and macrocells implemented in LTE bandwidths of 1 MHz, 5 MHz, 10 MHz, or 20 MHz. For outdoor small cells, 30-kHz subcarriers are suggested implemented into the TDD bands above 3 GHz, for example, Band 42 from 3.4 to 3.6 GHz and Band 43 from 3.6 to 3.8 GHz with 100-MHz or 80-MHz channel rasters. The 60-kHz subcarriers are suggested for indoor wideband implemented into the unlicensed band at 5 GHz using a 160-MHz channel raster. The 12-kHz subcarriers are suggested for Ka-band at 28 GHz on a 500-MHz channel raster. The 240-kHz and 480-kHz subcarriers are specified for future use.

5G and satellite standards are covered in more detail in Chapter 10, but it can be seen that there is a clear expectation that channel bandwidths need to scale from the present LTE 10 MHz implemented in passbands below 2 GHz to 500 MHz at Ka-band and in the longer term to 1 and 2 GHz in V-band and W-band. Note that the NEWLEO filings with the FCC, for example, the July

Table 2.5
Alignment of 3GPP Standards Process and WRC 2019

Phase A	Phase B	Phase C
2012	2013	2016
Releases 10 and 11	Releases 12 and 13	Releases 14, 15, 16
WRC 12	WRC 15	WRC 19

Source: 3GPP.

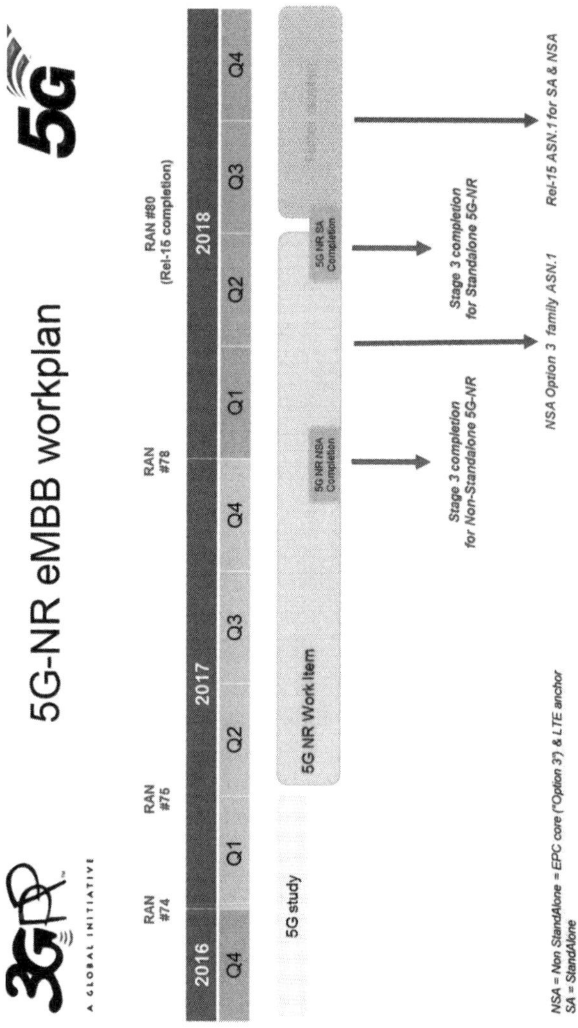

Figure 2.3 5G New Radio (NR) eMBB work plan. (Courtesy of 3GPP.)

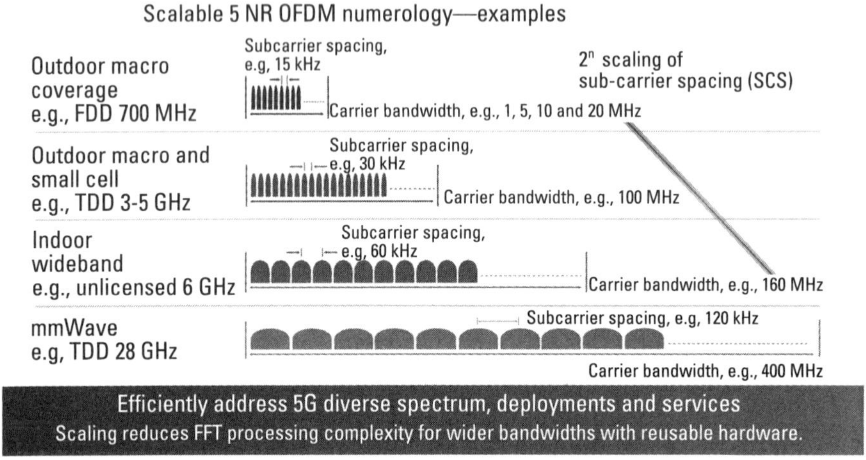

Figure 2.4 OFDM Numerology Image. (© 2017 Qualcomm Technologies, Inc. and/or its affiliated companies.)

2016 OneWeb filing, assume 500-MHz passbands for the user links in Ku-band (12.2–12.7 GHz uplink, 14.0–14.5 GHz downlink) and 500-MHz passbands for the feeder/gateway links in Ka-band (19.7–20.2 GHz downlink and 29.5–30 GHz uplink). The assumption is that the uplink channels are implemented as 125-MHz carriers and the downlink as 250-MHz carriers. User and IoT devices demodulate all traffic on each 250-MHz downlink channel and then discard the packets with headers that are not addressed to them. Note that for reasons of power efficiency, satellite systems do not use OFDM or quadrature amplitude modulation (QAM) but implement relatively simple amplitude phase shift keying (APSK). This has constrained AM components and therefore requires less (power consuming) linearity from the RF power amplifier. It could be argued that APSK is less spectrally efficient than QAM and OFDM but in practice spectral efficiency is achieved through spatial separation and polarization diversity. Figure 2.5 compares the two modulation types.

Considerable work still needs to be done on the merits/demerits of cosharing terrestrial and space spectrum with different physical-layer specifications. We revisit this topic in more detail in Chapter 10.

2.10 Regional Spectrum Policy

In terms of regional policy, there are some notable differences between the FCC and present ITU policy specifically around the 28-GHz, 38-GHz, and 39-GHz bands. In particular, the FCC is taking a robust approach to allocating primary access rights to mobile broadband operators at 28 GHz and 39 GHz in response to lobbying from AT&T and Verizon. Unsurprisingly, the satellite industry is

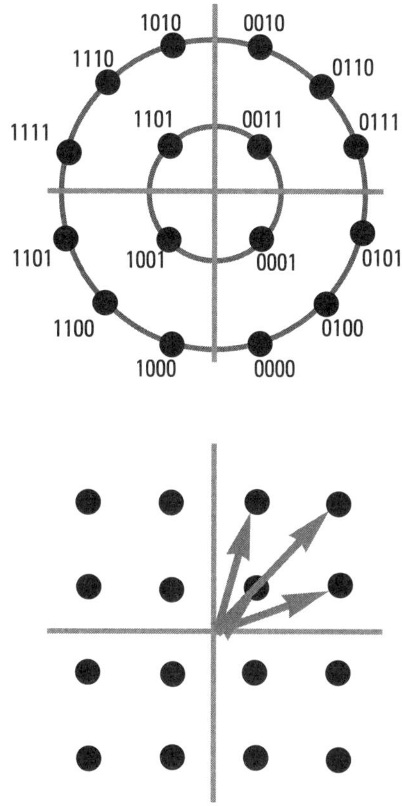

Figure 2.5 APSK (top) and QAM (bottom) modulation. (Thanks to Radio Electronics.)

objecting to this and can be expected to adopt a strong advocacy position for more limited access rights on a coshared basis at WRC 2019. This includes financial modeling, which suggests that mobile operator terrestrial deployment in the centimeter and millimeter bands could have a negative rather than positive impact on mobile operator earnings before interest, taxes, depreciation, and amortization (EBITDA) and enterprise value.

3GPP Release 15 focuses on sub-6-GHz 5G deployment (Table 2.6). The satellite industry could logically take this argument a stage further by arguing the case for 5G deployment in existing LTE spectrum including the new Band 71 at 600 MHz rather than in existing satellite bands. Alternatively (as we argue elsewhere in this book), it could be regarded as logical to provide access to these subgigahertz bands for satellite use.

There is potentially 155 + 155 MHz of low band spectrum available from 450 MHz to 900 MHz. This might seem modest compared to having access to several gigahertz of Ku-band, K-band, and Ka-band spectrum but this is quality spectrum, unaffected by weather, with favorable propagation, minimal

Table 2.6
Sub-6-GHz Band Allocations for Terrestrial 4G and 5G Spectrum

450		**600**			**700**			**800**			**900**		**Total**
Band 31 Brazil		Band 71 U.S.		Band	Band 28 APT		Band	E850 USA	Band 26	Band	Band 8 Europe and Asia		MHz
452	462	617	663		703	758		814	859		800	925	
457	467	652	698		748	803		849	894		915	960	
5	5	35	35		45	45		35	35		35	35	155+155
				12	696	729	20	791	832	U.S. ISM	902		
					716	746		821	862		928		
				17	704	777							
				AT&T	716	787							
				13	746	777							
				Verizon	756	787							
				14 First Net	758	788							
					768	798							

surface absorption, and scatter. From a regulatory perspective, the spectrum comes with clearly defined and highly protected access rights.

2.11 5G and Satellite at UHF

UHF could therefore provide a low-cost, relatively high data rate 5G connectivity option. It will be important for the satellite operators to show that they can compete with these potentially enhanced sparse network terrestrial options both in terms of price, throughput, and coverage particularly given the scale economy gains that can be realized by the Chinese vendor community amortized across their local high-volume 4G and 5 markets where base station shipments are counted or will be counted in millions of units, with user and IoT device shipments counted in billions of units.

From a cellular site perspective, it is hard but not impossible to implement smart antennas at these wavelengths, the challenge is to deliver performance gain within a 0.3-m-wide envelope panel antenna (one column of elements) to meet weight and wind-loading constraints. If the spectrum was more generally shared with the satellite community, then coexistence issues would need to be addressed. A large UHF array in space would not have a wind-loading issue.

2.12 5G in Refarmed Spectrum

A friend visiting Australia this year was surprised that his admittedly elderly phone did not work. The reason was that Telstra has turned off their GSM network.

With operators beginning to decommission their 2G GSM and 3G networks, it becomes at least theoretically possible to implement 5G into any 4G bands anywhere from Band 31 (450 MHz) to Bands 42 and 43 (3.4–3.8 GHz).

However, as we pointed out in Chapter 1, it is not only the available bandwidth that is important but also the bandwidth ratios (the ratio of the passband bandwidth to the center frequency of operation). This means that there is no obvious home for contiguous 200-MHz 5G channels below 3 GHz even if they could be supported through a traditional acoustic filter chain in the front end of a user or IoT device, which seems unlikely.

Additionally, not all operators will want or need to decommission their GSM networks, particularly if significant GPRS vertical market user groups need to continue to be supported. Release 13 also introduced Enhanced Coverage GSM (EC-GSM) with additional channel coding, which could potentially be a cost-effective option for some deep rural areas. It is therefore not immediately apparent what gains could be realized from 5G in refarmed spectrum over and above LTE Advanced and LTE Pro and enhanced legacy technology options such as EC-GSM.

An additional option is to implement 5G in discontinuous channel aggregated spectrum, but presently there are so many operator specific band plan options that it seems unlikely that any global scale economy can be achieved for 4G let alone 5G user and IoT devices.

This topic is covered in greater detail in *5G Spectrum and Standards* including background on some of the performance trade-offs implicit in supporting high bandwidth ratios and or aggregated channels. In summary, it is possible to design front-end RF architectures that can process multiple existing RF bands in parallel to achieve high headline data rates. The assumption is that the ability to send data quickly will reduce power drain, but this has to be set off against lower RF efficiency and physical layer clock processor overheads. It is difficult to design a front-end architecture that is good for processing multiple and single bands and therefore a user could find a device that delivers high headline data rates might, for example, perform less well at a cell edge (low carrier to interference) or in marginal coverage areas (low signal to noise). Traditionally, these apparently rather prosaic user and IoT device RF performance compromises tend to be overlooked in physical layer design and network economic modeling.

2.13 The FCC, the ITU, and Sovereign Nation Regulation: Similarities and Differences Between Terrestrial and Nonterrestrial Networks

This brings us reasonably neatly to a discussion on the differences that exist region to region and country to country in terms of how spectrum is allocated, auctioned, and regulated and the commonalities and differences between terrestrial spectrum management and space spectrum management.

The first obvious difference is that satellite systems are servicing users from space. Strictly speaking, nonterrestrial systems also include LTE Air to Ground used, for example, to provide two-way communications with helicopters. Studies have also been made using low-cost drones to provide on-demand coverage for emergency response and disaster relief.

Figure 2.6 shows an example of air-to-ground LTE and an LTE pico base station being flown on a drone in a Verizon test.

If this becomes at all common it implies a need to reconfigure terminals and their antennas to receive signals vertically from above rather than on a more or less horizontal or at a low elevation angle. If 5G air-to-ground or drone-based 5G is deployed in the same bands as 5G NGSO and GSO satellites, then it can be considered either as a problem (mutual interference) or an opportunity

Figure 2.6 LTE air to ground and a drone-based LTE base station. (With thanks to American Aerospace Technologies, Incorporated and Verizon Wireless.)

(shared channel bandwidth between 5G and NGSO/GSO networks separated in terms of vertical and horizontal signal). Note that the Verizon tests to date have been 4G LTE rather than 5G.

2.14 Air to Ground for Public Protection and Disaster Relief: AT&T FirstNet, BT EE, and the Australian NBN as Examples of LTE and Longer-Term 5G Emergency Service Radio Networks

These instant LTE networks in the sky are important for public protection and disaster relief (PPDR) incident response and have to be part of a mobile broadband operators network offer if they bid for and win public safety radio contracts. Examples include the AT&T FirstNet network tasked with the replacement of 10,000 separate legacy radio systems in the United States, BT EE replacing the Airwave TETRA network in the United Kingdom, and the Australian National Broadband Network in Australia. The U.S. network requirement was specified after September 11, 2001, and includes coverage into public buildings, shopping concourses, and underground areas.

These networks require geographic coverage including coverage in deep rural areas and hard-to-reach urban and in-building locations specified as a service level agreement. These can be addressed to an extent by network buildout and by equipping emergency vehicles with LTE base stations. The Australian NBN also includes two GSO satellites in the service to meet rural coverage and network resiliency requirements.

2.15 GSO and NGSO Terminology

From a regulatory viewpoint, satellite systems are generally characterized as being either GSO (geostationary) or non-GSO (NGSO). NGSO includes MEO and LEO satellites and any satellite that appears from the ground to be moving. GSO systems are obviously also moving but at the same speed as the Earth's rotation and therefore appear stationary as seen from Earth. Because of their fixed orbital position above the Earth, a GSO constellation (a number of GSO satellites) can clearly be dealt with on a regional or country-by-country basis; in addition, interference issues are generally related to fixed entities and are reasonably easy to manage.

In contrast, NGSO systems, for example, LEO satellites, overfly many regions and countries requiring them to be compliant with many (and potentially various) different regulatory regimes in order for them to be allowed to deliver service to users. Interaction and interference with GSO systems also needs to be managed, with GSO systems taking the higher ground (in more ways than one). In this book, we use the term NGSO when referring to regulatory issues, although to all intents and purposes NGSO and LEO are interchangeable in terms of the actual systems being discussed.

In order to get a satellite project literally off the ground, there are a number of initial hurdles that have to be overcome. The NEWLEO entities including OneWeb and Space X have dealt with some but not all of these.

For U.S. companies in particular, the process generally starts with a filing submission to the FCC as the United States is still the largest and most influential sovereign entity in the global satellite sector, although China and India are catching up quickly.

Every sovereign nation in the world has the right to determine how radio spectrum is used in and theoretically above its territory and, in particular, a right to demand that particular and occasionally country specific coexistence conditions are applied.

The World Trade Organization, within the framework of the General Agreement on Trade in Services (GATS) while recognizing this sovereign right of States to manage the frequency spectrum in terms of their own objectives, works to develop the instruments required so that exercise of that right does not result in barriers to trade in services between its members. In this context, the establishment of standards at regional and global levels facilitates efficient and economical use of the spectrum and the development of radio services. The ITU [11] works in parallel with the WTO to provide a regional framework that allows sovereign nations to submit and discuss their spectrum requirements at regional level (Regions 1, 2, and 3). The ITU Regions (Figure 2.7) are specified in Articles 5.2 to 5.22 of the ITU Radio Regulations.

The outputs from these regional meetings then go forward into the World Radio Conference (WRC) process. The last WRC took place in 2015; the next will take place in November 2019. These meetings are enormous with typically 7,000 delegates with a flag system implemented in plenary sessions to adjudicate national sovereign representation.

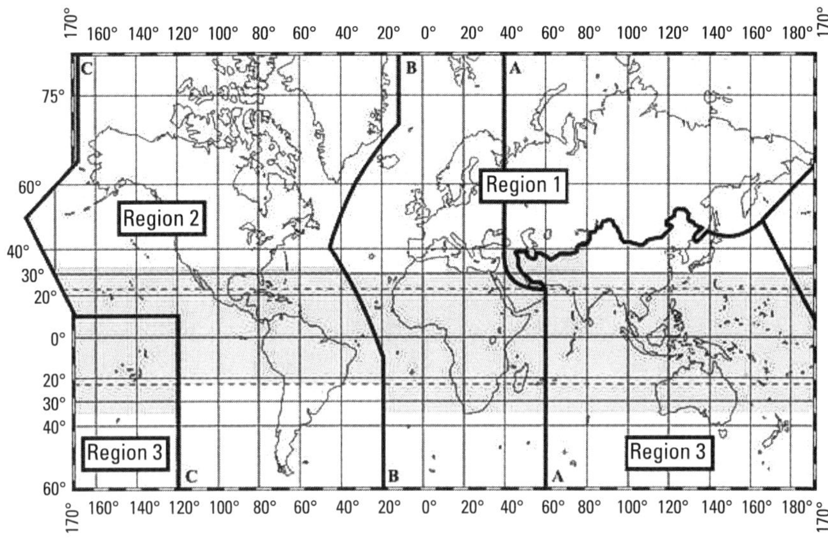

Figure 2.7 ITU regions.

2.16 Why Country and Regional Differences Are Important for Global Connectivity

One of the more compelling reasons to consider satellite systems for global connectivity is that potentially a car, truck, bus, ship, plane, train, or other large or small moving or static object could be shipped to any country in the world and any place in any country, and be seamlessly and continuously connected ideally through one integrated global network.

This is already the case for example with the Orbcomm VHF services provided to John Deere, Volvo, Caterpillar, and Hitachi Construction and from users of the Iridium system. However, these are relatively narrowband systems (1 + 1 MHz at VHF for Orbcomm and 10 + 10 MHz at L-band for Iridium), with clear and well-documented access rights across the world.

By comparison, OneWeb and Space X and other NEWLEO contenders propose to deploy a wideband radio system with a 2-GHz downlink passband and 1-GHz uplink passband in Ku-band and are sharing (sometimes referred to as cosharing spectrum) that spectrum with TV broadcast, video distribution multichannel video and data distribution service (MVDSS), and TV and internet delivery technology licensed for use in the United States by the FCC; this terrestrial-based wireless transmission method reuses direct broadcast satellite (DBS) frequencies for distribution of multichannel video and data over large distances, GSO and MEO satellite systems, and other two-way communications systems including military radio, deep space radio, and radio astronomy. The gateway links are deployed in K-band and Ka-band and have similar coexistence issues that have to be addressed.

Critically, there are regional and sometimes country-specific differences in the way that spectrum band plans and radio system technologies are deployed, with the result that in-band and out-of-band (OOB) emission requirements can be significantly different on a regional and country basis.

The established satellite operators (legacy SATs) have had technical and regulatory teams addressing these differences in some cases for the past 50 years. Iridium and Globalstar, the "new kids on the block," have more than 20 years of experience. The NEWLEOs have to deal with this regulatory complexity in a compressed time scale (deployment by 2019/2020) to meet FCC requirements. To an extent, this can be achieved by recruiting regulatory capability.

The ITU specifies that GEO satellites have priority over LEO satellites with regard to frequency usage. The problem is that the NEWLEO satellites will be regularly passing between users on the ground and GEO and MEO satellites, while using the same Ku/K-band and Ka-band frequencies.

So it is important to understand the particular intersystem interference mitigation measures proposed by the NEWLEOs. Generally, these tend to be documented in FCC filings, a consequence of the historic market dominance of the U.S. satellite industry and consequent regulatory influence of the FCC.

We recommend readers study the original OneWeb 2016 FCC filing as a starting point [12]. Interference mitigation measures and the software models associated with them need to be agreed by the ITU and can be challenged by other entities including incumbent operators' sharing this spectrum.

The OneWeb filing referenced above is based on a proposal for 720 satellites. Recent press statements from OneWeb [13] indicate that they have production options for 2,000 satellites. Adding satellites to a constellation increases capacity but also increases flux density if the same power levels are used on each satellite. OneWeb will have to demonstrate that a higher count constellation will still conform with EIRP and flux density limits (see Chapter 7) and provide guidance on how this might affect the number of ground (Earth gateway) stations needed and their likely location and composite uplink and downlink power. On the positive side, increasing the number of satellites in the constellation increases the number of times a satellite will be directly overhead with maximum vertical separation from other satellite systems at a lower elevation or terrestrial 5G with close to horizontal elevation.

Like other ITU filings, the rights to use these frequencies for an NGSO are granted on a first-come, first-served basis. As there are multiple NGSOs in the planning and implementation stage (OneWeb, LeoSat, SpaceX, Telesat), progress is dependent on the seniority of the entities in the filing process. Similarly, the interference and protection ratios need to be modeled for the composite interference produced by all proposed constellations.

2.17 RF Power and Interference

Satellite operators are licensed to operate in defined frequency bands with defined maximum (transmit) power levels, and with conditions applied relating to interference with other systems. Transmit power is normally specified as effective isotropic radiated power (EIRP), the measured radiated power in a single direction. The result of this transmitted power, the design (polar response) of the transmit antenna on the satellite, and the orientation of the satellite relative to the receiver on the ground (i.e., overhead or at a glancing angle) is the power flux density (PFD). For a ground-based user terminal, the flux density, when combined with any antenna gain, will dictate the signal level at the receiver input.

Interference at a receiver caused by other transmitters may be in-band or out-of-band. A protection ratio may be specified, which defines the minimum value of the wanted-to-unwanted signal ratio, usually expressed in decibels at the receiver input, to achieve a specific reception quality (e.g., bit error rate and throughput).

2.18 The Importance of Intersatellite Switching

The 2016 OneWeb filing [14] identifies a need for 50 ground stations and additional stations at high latitude to support telemetry and control and to manage through-life maintenance, orbit-keeping, and end-of-life deorbiting. Securing licensing and landing rights for the gateways in 50+ locations around the globe will be challenging for OneWeb and other NEWLEO operators and may be the dominant pacing issue for revenue operations and global deployment. Increasing the satellite count from 720 to 2,000, which is presently proposed by OneWeb [15] may require more gateways, although this is not stated. A gateway is essentially an antenna farm with multiple dishes (of about 2.5-m diameter) pointing at different parts of the sky and the higher count could presumably be supported by more antennas per site, but the site will get bigger. This may seem a trivial point, but someone has to find suitable sites, purchase or lease land, and arrange planning permission across a range of different planning regimes (and source electrical power and backhaul connectivity). The alternative is to uplink to a MEO or GSO and then return via the GSO downlink back to Earth. This is used in military radio systems (and Hubble and the International Space Station) but introduces additional latency. However, it would produce substantial cost savings.

2.19 Landing Rights

We have said that satellite operator assets can be summarized as spectrum access rights, orbital rights, and landing rights (permission to deliver services into and out of a sovereign country) but also gateway assets.

The technical issue with landing rights is that it requires RF power to be focused on a country or region from a satellite or satellites either over flying the land mass (NGSO) or always visible at a fixed inclination angle (GSO). User devices and ground stations will also be transmitting on the uplink.

If a sovereign country considers that existing satellite systems or terrestrial systems including military satellite and terrestrial radio or satellite TV receivers could be compromised by a newly proposed service, then they can request and insist that spot beams from the satellite are turned off or that RF output power is reduced. Therefore, there may be countries in which OneWeb and other NEWLEO operators cannot provide coverage or can only provide coverage at lower RF output power.

2.20 Interference Management

The additional mechanism used by OneWeb and all other proposed high throughput Ku-band, K-band, or Ka-band NGSO systems to meet country-specific EIRP and flux density limits is angular power separation (see Chapter 6 for alternative methods for cosharing and interference management). Essentially, this means that at high latitudes, the assumption is that there will nearly always be a LEO nearly overhead delivering RF energy via a spot beam to a group of geographically proximate users within a cell. A car, for example, will be demodulating 250 MHz of channel bandwidth, which will have a number of users sharing the bandwidth. The traffic of interest is identified from the Transmission Control Protocol (TCP)/Internet Protocol (IP) packet headers.

Note in passing that the contention ratio will have a direct impact on the available bandwidth. This is an important consideration and needs to be included in network test plans. On the uplink, a similar time division multiplex (TDM) contention protocol is used across a 125-MHz channel bandwidth subdivided into narrower (<20 MHz) channels. Contention rates are an important parameter that requires careful specification in service-level agreements.

The basis for frequency sharing is that terrestrial systems or GSO systems will be receiving RF energy from a much lower elevation angle (the satellite TV dish on the side of your house being a good example) and therefore system cross-talk will be minimized. We cover this in more detail in the next section.

Conversely, nearer the equator, GSO satellites will be shining directly downwards. To avoid interference, OneWeb and other new LEOs use a technique known as progressive pitch which means that RF energy is delivered at an inclined angle from satellites either side of the equator rather than above the equator. RF power is then turned down or off as the satellite moves across the cone of visibility of the victim receiver with service delivered from a satellite nearer the horizon. Progressive pitch can also be achieved by altering the pitch of the satellites as they traverse over the equator. This is achieved by using reaction wheels [16], standard fitment to satellites to alter spacecraft orientation, also known as momentum wheels (a major supplier is Blue Canyon Tech [17]), to establish the spin rate and direction on each equatorial traverse (every 55 minutes).

The impact of this on user links needs to be considered. For moving objects like cars, trucks, or buses, a key reason to use high-count LEO constellations is that they are nearly always nearly overhead, which minimizes blocking from buildings and trees. This advantage would potentially disappear due to the need to meet country-specific EIRP and flux limits by using a low elevation angle. Note that this will also result in a longer path length, which will increase atmospheric fading and require a higher rain fade margin and increase path link delay.

This would be potentially a problem in equatorial countries with tall buildings, Singapore being one example.

Thus, it can be seen that detailed country-specific regulatory requirements, with the implied need to meet EIRP and flux density limits determined by angular power of arrival and departure could have an impact on service availability and service quality and could mean that the user experience could be variable from market to market and occasionally unavailable.

On a more positive note, it can be seen that NEWLEOs could be a useful complement to terrestrial 5G particularly for terrestrial 5G implemented in Ku-, K-, or Ka-band where building and surface scatter absorption will be significant. In particular the angular separation between the nearly always nearly overhead LEO signal at higher latitudes and the signal energy coming in at effectively a 90° offset suggests opportunities for in-band frequency reuse, particularly if polarization diversity is also used. How well this works will be a function of the antenna design, a topic that we tackle in Chapter 6.

2.21 Spectrum Access Rights

Spectrum access rights are closely analogous to property rights and the regulatory and legal frameworks are similar for satellites and terrestrial systems although NGSO satellites are more complex because they are moving.

Spectral access rights can be either primary access, coprimary access or secondary access with primary access being implicitly the most valuable asset, as seen in Table 2.7.

Table 2.7
Spectrum Access Rights

Primary Access	Guaranteed sole usage and protection from interference (including the ability to stop competitors deploying systems on the basis that they might cause interference rather than waiting for the interference to happen and be detected and measured)
Coprimary Access	Agreed shared usage by 2 or more operators based on enforceable technical (coexistence) standards
Secondary Access	Usage allowed on a secondary basis, must accept interference from other users (who must themselves comply with agreed power limits)

By definition, with LEO NGSO we are talking about shared access regimes where the existing (GSO) incumbents have well established existing primary access rights.

Regulators judge a newly proposed service on the basis of its potential economic value (impact on national, regional, or global GDP), social value (for example, bridging the digital divide) and political value (satellite TV being a prime example).

Mr. Greg Wyler, the founder of OneWeb, has proved adept at playing this regulatory game of poker. His previous company, O3b, acquired spectrum from Teledesic, a high-count Ka-band LEO constellation, which ran out of money in 2002, having absorbed the best part of $1 billion of Mr. Craig McCaw's considerable fortune (Figure 2.8).

The spectrum is divided into subbands with designated equivalent power flux density limits (EPFD) and bands where there are no limits but where interference has to be coordinated with GSO operators.

The FCC submission was for a MEO constellation and the stated market/business model was connecting the unconnected other 3 billion, hence the name.

O3b inconveniently had to raise capital in 2008, the year that Lehmann Brothers went bankrupt [18], and it is a tribute to the persuasive skills of the Wyler management team that the constellation launched and more or less met its business plan objectives.

However, it achieved this by substantially altering the market focus of the business that now supplies internet connectivity to cruise ships 40° either side of the equator. The average cruise ship now consumes in the region of 500 Mbps of internet bandwidth in peak hours, a highly profitable market (although O3b claims it also provides service to some parts of the Amazon and the Pacific Islands). Cruise ships have the advantage that they operate for most of the time outside sovereign jurisdiction, meaning that O3b could avoid the whole pesky business of negotiating country-by-country landing rights. This illustrates that serviced markets can change substantially from the initial FCC filing.

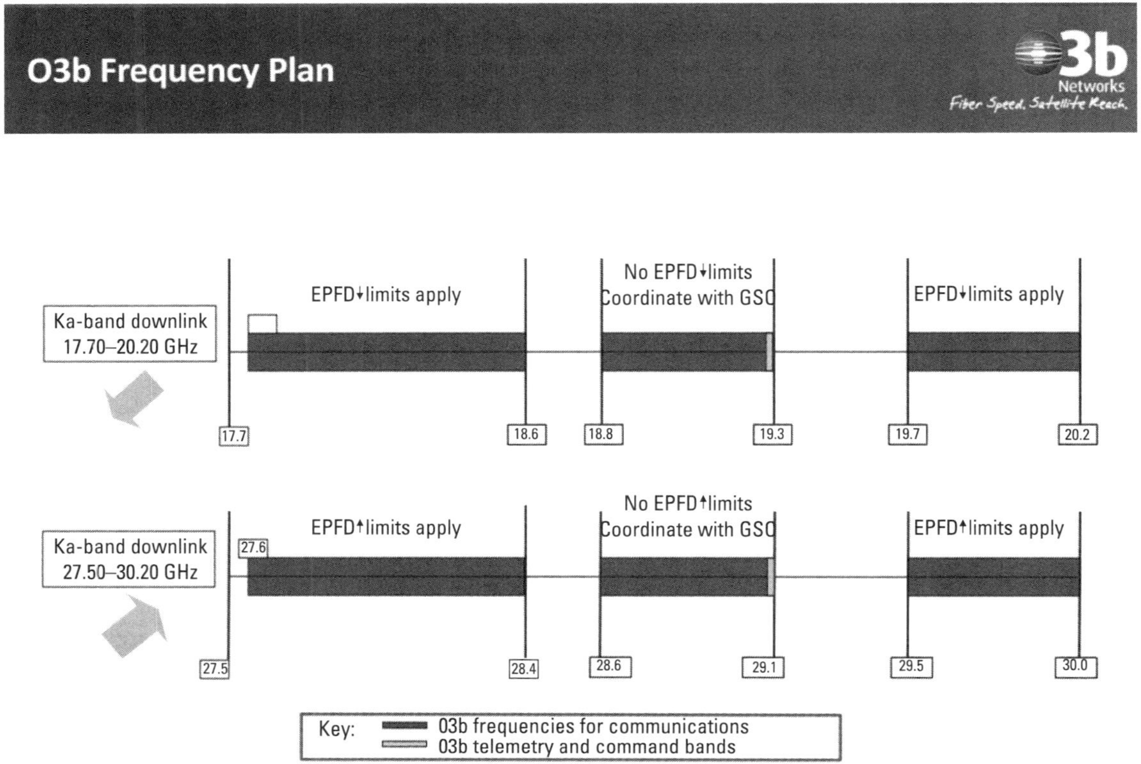

Figure 2.8 O3b frequency plan, formerly the band plan for Teledesic. (Thanks to SES Networks.)

Mr. Wyler left O3b in 2012 and founded OneWeb initially called World-Vu. The entity also uses the alternate name L5 in regulatory filings with the ITU and is registered in the United States and Jersey.

In March 2017, OneWeb submitted a filing with the FCC for an additional 2,000 satellite constellations in V-band (40–75 GHz), although at this stage this must be considered an essentially speculative move. Space X and Boeing and a number of other potential new LEO entities have also submitted V-band constellation proposals.

OneWeb managed to acquire the spectrum and access rights owned originally by Skybridge Inc. [19], a U.S. entity established in the 1990s to roll out a high satellite count LEO constellation in Ku-band (user uplinks and downlinks) and Ka-band (gateway uplinks and downlinks). Skybridge went into administration before the constellation could be realized (Figure 2.9).

The passbands are shown below with designated access rights from the FCC for the United States with a listing of other entities sharing the spectrum including mobile and fixed services and broadcast services.

Figure 2.10 shows the K-band and Ka-band spectrum access rights.

The OneWeb submission closely follows this band plan with some minor amendments. The band plan submission is shown in Table 2.8.

Although OneWeb satellites have the capability to operate in the Earth to space direction in the 12.75–13.25-GHz band and the space-to-Earth direction in the 19.7–20.2-GHz band, FCC authorization is not being requested for these bands and they will not be used in any U.S. territories. OneWeb

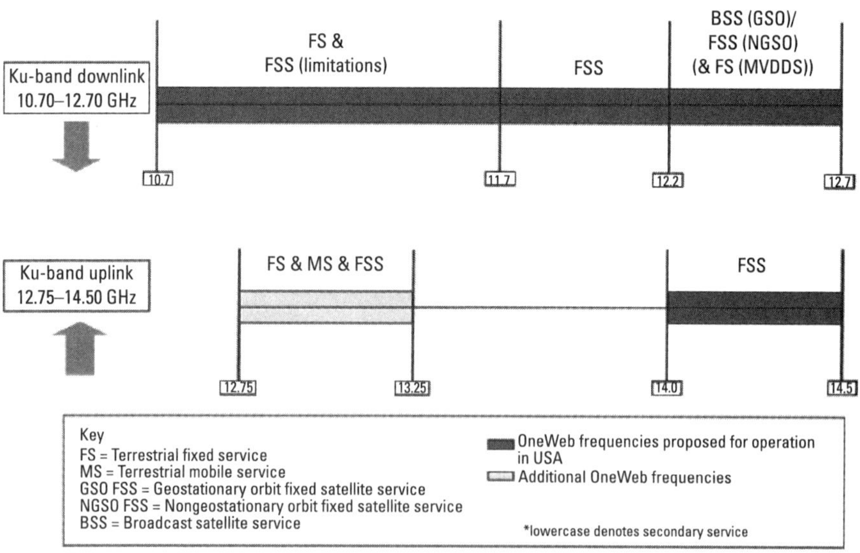

Figure 2.9 Ku-band spectrum rights acquired from Skybridge by OneWeb (from the July 2016 FCC filing).

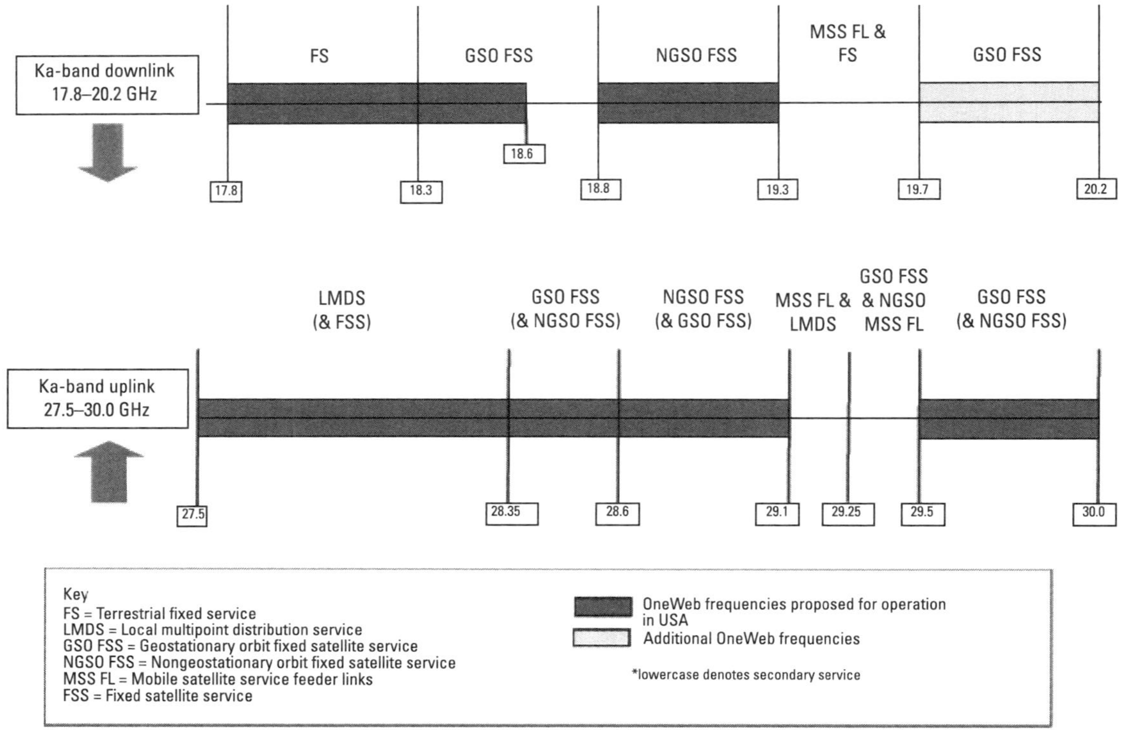

Figure 2.10 Ka-band spectrum access rights acquired from Skybridge by OneWeb (from the FCC OneWeb filing in July 2016).

Table 2.8
OneWeb FCC Submission in July 2016

Link and Direction	Frequency	Frequency
Gateway to satellite	27.5 GHz, 29.5 GHz	29.1 GHz, 30.0 GHz
Satellite to sateway	17.8 GHz, 18.8 GHz, 19.7 GHz	18.6 GHz, 19.3 GHz, 20.2 GHz
User/IoT device to satellite	12.75 GHz, 14.00 GHz	13.25 GHz, 14.5 GHz
Satellite to user/IoT device	10.7 GHz	12.7 GHz

committed in the filing to providing the FCC with the requisite deployment time scales for satellite deployment in accordance with FCC Article 25.118(f).

2.22 NGSO to GSO Interference Mitigation

Skybridge had progressive pitch angular power separation in their original FCC filing submissions as a mechanism for meeting U.S. Ku-band EIRP and flux density limits and protection ratios to the shared services supported in and adjacent to the passband. The FCC was subjected to significant lobbying from these entities sharing the spectrum who questioned the validity of the models used to calculate interference levels and the efficacy of the proposed mitigation measures. This continues today, with a present example being the MVDDA/MVDS coalition representing companies such as Echostar/Dish Networks, which are deploying, have deployed, or propose to deploy multichannel video distribution services and/or 5G services [20, 21]. Space X is similarly challenging the interference models presently used by the FCC on the basis that they were developed prior to and for the WRC 2000 meeting and fail to take into account the dynamic interference capabilities of high satellite count LEO constellations based on progressive pitch and power control.

Heading off these technical and legal challenges and advancing these technically complex coexistence arguments will be an ongoing and onerous task for OneWeb and other new LEO GSO entities. The legal and litigation process will absorb management time and money and may delay deployment in the United States and other global markets.

An FCC fact sheet produced in June 2017 summarized the obligations that OneWeb will have to meet in order to deploy a network in the proposed passbands (Figure 2.11). This includes coexistence with GSO operators, other NGSO operators, terrestrial operators, upper microwave flexible use service, and operators in the 17.8–18.6 GHz where OneWeb will only be only permitted to operate on a noninterference, nonprotected basis.

Note that this is described as market access rather than spectral access rights. The access grant ruling is only applicable to the U.S. market. This is

June 1, 2017

FCC FACT SHEET*
OneWeb Market Access Grant
Order and Declaratory Ruling - IBFS File No. SAT-LOI-20160428-00041

- **Background**: OneWeb is seeking access to the U.S. market for a proposed non-geostationary-satellite orbit (NGSO) fixed-satellite service (FSS) satellite system, consisting of 720 satellites distributed across 18 near-polar orbital planes at an altitude of approximately 1,200 kilometers. The proposed grant would be the first Commission approval to facilitate a new generation of NGSO FSS large satellite constellations proposing to provide ubiquitous low latency broadband connectivity across the United States, including some of the most remote areas in places like Alaska where broadband access has not been possible before.

What the Order Would Do:

- Grant OneWeb's request for a declaratory ruling concerning the conditions under which it will be permitted to provide broadband communications services with its NGSO FSS constellation to the United States using frequencies in the Ku- and Ka-bands, specifically the 10.7-12.7 GHz, 14-14.5 GHz, 17.8-18.6 GHz, 18.8-19.3 GHz, 27.5-29.1 GHz, and 29.5-30 GHz frequency bands. As such, the Order provides a blueprint for the earth station licenses that OneWeb or its business partners must obtain before providing service in the United States.
- Specify conditions intended to protect or accommodate other operations, including:
    - Geostationary-satellite orbit (GSO) operations: OneWeb operations will protect GSO operations by meeting equivalent power-flux density limits.
    - Non-geostationary orbit operations: OneWeb operations will comply with the avoidance of in-line interference spectrum sharing method specified in 47 CFR § 25.261(b)-(d) with respect to any NGSO system licensed or granted U.S. market access pursuant to the processing round initiated in Public Notice, DA 16-804.
    - Terrestrial operations: OneWeb will protect terrestrial operations by meeting power-flux density (PFD) limits.
    - Upper Microwave Flexible Use Service (UMFUS): OneWeb operations will protect UMFUS operations in the 27.5-28.35 GHz frequency band in accordance with the rules adopted in FCC 16-89.
    - Operators in the 17.8-18.6 GHz Frequency Band: OneWeb operations will be authorized in this band only on a non-interference, non-protected basis.
- Require modification of OneWeb operations to bring them into accordance with any future rules or policies adopted by the Commission.

Figure 2.11 OneWeb market access grant fact sheet. (Reproduced with permission of the FCC.)

therefore not substantially different from terrestrial mobile spectrum access rights, which also have to be negotiated and bid for on a country-by-country basis, although note that these satellite bands have not historically been auctioned but are made available in return for specific service obligations including geographic coverage requirements.

2.23 FirstNet and the 2012 Spectrum Act

One analogy to this would be the AT&T agreement with FirstNet and the U.S. government and the 2012 Spectrum Act determining that $7 billion would be allocated to fund network construction (see Section 2.14). AT&T has draw-down rights on this construction subsidy budget together with access to 20

MHz of 700-MHz spectrum although with onerous service obligations attached. A similar regulatory approach to Ku-band, K-band, and Ka-band spectrum could potentially involve incentives to provide fiber equivalent access to remote rural communities or very high throughput satellite (VHTS) data rates for emergency services response in remote outdoor locations. The NEWLEOs might be in a good position to provide additional coverage for first responder user groups presently being supported on Release 8 LTE.

2.24 Fiber Access and Wireless Access Rights

Incidentally, the close down of Teledesic in 2002 continues to make waves, some of them positive. The fiber assets of the holding company XO Communications (formerly Next Link), were recently purchased by Verizon for $1.8 billion with an option to acquire some of the residual spectrum access rights of the company at 28 GHz [22].

In April 2017, AT&T paid $1.25 billion to acquire the access rights of Straight Path at 28 GHz and 39 GHz. Next Link and Straight Path are both examples of entities set up to establish local multipoint distribution services (LMDS), but coverage rights do not extend to all or indeed many of the 289 cellular market areas designated by the FCC and there is a particular lack of deep rural coverage. These were essentially speculative spectrum acquisitions by companies with limited engineering resource or network rollout experience, and it might be argued that they were really only set up to be bought out by Verizon or AT&T. Note that the LMDS license conditions specify fixed but not mobile services [23].

This implies a need for line of sight between the base station/access point and user/IoT fixed terminal or customer premises equipment (CPE) to avoid the high scatter losses and surface absorption losses at these shorter wavelengths/higher frequencies.

This is hard to realize both in urban and rural areas and involves insupportably high real estate costs, particularly if potential new operators do not have tower or building assets. Operators such as AT&T and Verizon can at least build out from their existing cellular and backhaul infrastructure and site assets. Even bearing this in mind, it can be seen that there are some persuasive arguments in favor of direct line of sight from above, the NEWLEO nearly always nearly overhead (NANO) access model. Note that GSO coverage at higher latitudes will be at a low elevation angle and will therefore suffer from blocking from buildings and foliage and from surface scatter (similar to the terrestrial propagation model). Conversely, NEWLEO elevation over the equator will need to be inclined in order to meet GSO protection ratios and will similarly suffer from blocking. Note that wet foliage will have a higher absorption loss and it rains a lot in the tropics.

The best option is to combine LEO, MEO, and GSO footprints to deliver always overhead downlink and uplink visibility. This would potentially fill pretty much all the coverage gaps, and more importantly would scale easily to other global markets. We revisit this topic in Chapter 7.

These U.S.-specific spectrum acquisitions explain why U.S. operators supported by the FCC are intent on developing 5G at 28 GHz and 39 GHz irrespective of the reservations and objections put forward by other operators in other sovereign countries.

2.25 Fixed Point-to-Point and Point-to-Multipoint Microwave Backhaul

LMDS, to all intents and purposes, was a U.S. regulatory construct designed to encourage new market entrants to provide an alternative to fiber in places where fiber operators did not want to go. As such, it depended on realizing a capital expenditure (CAPEX) and operational expenditure (OPEX) cost base lower than fiber on an actual and cost-per-bit basis, which was always going to be optimistic particularly given the failure of previous attempts to realize cost-effective fixed access wireless broadband.

We refer the reader to Chapters 9 and 10 in *5G Spectrum and Standards*, where we reviewed the spectrum and band plans used for terrestrial backhaul and microwave links. These are effectively identical in hardware terms to LMDS although with a different purpose (backhaul rather than internet broadband access to individual users and sites). Just to summarize, licensed link equipment at 28 GHz typically delivers 400-Mbps peak throughput through a 56-MHz channel with 38 dBi of isotropic gain through a dish antenna. A 38-GHz link with a 56-MHz aggregated channel supports 500 Mbps with 50 dBi of antenna gain. Spectrum at 42 GHz, 70 GHz, or 80 GHz uses 112-MHz or 250-MHz channel spacing with high-level modulation to deliver 1 Gbps. The 70-GHz and 80-GHz links can also achieve a headline 1-Gbps data rate by aggregating four 250-MHz channels together. The additional bandwidth means that lower-order modulation can be used.

Clearly, there are opportunities to realize scale economy benefits by reusing or cross-amortizing link hardware for more general point-to-multipoint and multipoint-to-multipoint networks. Satellites could also play a greater role in providing backhaul. The NEWLEO operators seem particularly confident that they can deliver backhaul at lower cost with adequate latency control.

2.26 Legacy LEO and GSO Operator Spectrum

OneWeb, Space X, and LeoSat (and Sky Space Global and Boeing) are between them producing a constant flow of announcements and proposals for NEW-LEO constellations. All of them are actively engaged in producing FCC filings and ITU submissions.

This occasionally frenetic activity should however be viewed in the context of ongoing upgrades by existing LEO operators Iridium and Globalstar and ongoing GSO upgrades. Iridium and Globalstar are engineering these upgrades within their existing L-band spectrum allocations. In December 2016, the FCC adopted rules permitting Globalstar to deploy a terrestrial low-power broadband network using 11.5 MHz of the company's 2.4-GHz (S-band) spectrum (2,483.5–2,500 MHz) to support small cell deployment for LTE networks. It utilizes a 22-MHz-wide Channel 14 in 2.4 GHz including 11.5 MHz on a licensed and 10.5 MHz on an unlicensed basis for Terrestrial Low Power Service (TPLS Wi-Fi). In contrast to Iridium, Globalstar does not use intersatellite switching (this is known as a bent pipe system). This reduces constellation cost and constellation complexity but also reduces end-to-end latency control. It also incurs additional ground station costs.

GSO constellation upgrades divide into Ku-band and Ka-band upgrades with the Ka upgrades generally described as high throughput satellite constellations. The Ka-band constellations (26–40 GHz) require a higher rain fade margin but can deliver more isotropic gain from the shorter wavelength fractional beamwidth antennas and will typically have anything between 12 and 100 spot beams. The passbands are channelized on 250-MHz channel rasters (similar to the proposed 3GPP Release 16 5G standard).

Table 2.9 shows one of the Inmarsat passbands. Compare this, for example, with the Iridium uplink and downlink in L-band of 10 by 10 MHz, a two-order of magnitude difference in available bandwidth.

2.27 V-Band and W-Band

At the time of this writing, Boeing and five other companies, SpaceX, OneWeb, Telesat, O3b Networks, and Theia Holdings [24], had all informed the FCC that they have plans to field constellations of V-band satellites in non-

Table 2.9
Inmarsat Ka Passbands with 250-MHz Channelization

	28.35 GHz	**29.60 GHz**	**29.25 GHz**		**30.0 GHz**
Polarization 1	250 MHz		250 MHz	250 MHz	250 MHz
Polarization 2	250 MHz		250 MHz	250 MHz	250 MHz

geosynchronous orbits to provide communications services in the United States and in the rest of the world's markets.

The FCC originally deferred on Boeing's request to operate between the 42-GHz and 42.5-GHz and 51.4-GHz to 52.4-GHz bands, but Boeing subsequently submitted a new application to the agency asking to use the 37.5-GHz to 42.5-GHz range of V-band for downlinking from spacecraft to terminals on Earth, and two other bands (47.2 to 50.2 GHz and 50.4 to 52.4 GHz) for uplinking back to the satellites. The company's proposed constellation would consist of 1,396 to 2,956 LEO satellites in 35 to 74 orbital planes at 1,200 km providing a footprint of thousands of 8–11 km cells. The industry rumor mill in 2017 suggested that Apple was providing finance or had at least expressed interest in financially supporting the Boeing V-band constellation [25].

Theia Holdings is a breakout company from the European Space Agency specializing in small CubeSats for communications and remote sensing. The other submissions describe their use of V-band spectrum as extensions to their Ku-, K-, and Ka- band proposals. SpaceX, for example, proposes a VLEO, or V-band LEO constellation of 7,518 satellites to follow the operator's initially proposed 4,425 Ka-band and Ku-band satellites. Canada-based Telesat describes its V-band LEO constellation as one that "will follow closely the design of the Ka-band LEO Constellation," using the same number of satellites as the initial proposal (117 satellites excluding spares) as a second-generation overlay. OneWeb informed the FCC that it wants to operate a subconstellation of 720 LEO V-band satellites at 1,200 km, and another constellation in MEO of 1,280 satellites. Added together, this would expand the OneWeb constellation by 2,000 satellites. OneWeb intends to dynamically assign traffic between the LEO and MEO V-band constellations based on service requirements and the data traffic within coverage areas. OneWeb's application for MEO V-band orbit and access rights follows the Viasat submission for 24 MEO satellites to augment their existing Viasat 3 constellation based on the companies' 3-Tbps-throughput satellites currently being planned and built or rather, financed. Viasat coupled its submission for the use of V-band with its application for the MEO Ka-band orbit and access rights. O3b told the FCC that it wants market access to V-band for up to 24 additional satellites that would operate in a circular equatorial orbit as a constellation called O3bN.

2.28 Summary

Many of the established GSO operators have over 50 years of experience delivering broadcast and two-way communication services to terrestrial customers including consumers, government agencies, the military and industry and business sectors. Even when privatized, they are still regarded as critical national assets and their spectral access rights are conferred accordingly.

There are now two NGSO operators, Iridium and Globalstar, that have been providing connectivity from their polar orbit LEO constellations for over 20 years. Their customers include government agencies, the military, public protection and disaster relief agencies, mining and exploration industries, and anyone who needs connectivity more or less anywhere at any time. Although both constellations required refinancing, they are presently in good technical shape and undergoing substantial constellation upgrades.

GSO and NGSO satellites have also been used for many years for Earth sensing and imaging. The paths of successive hurricanes in 2017 were tracked with exquisite precision from space. GSO and NGSO satellites are also used to track moving objects including airplanes, ships, and the odd missile heading towards Japan. These safety-critical, life-critical system requirements are reflected in protection ratios designed to ensure that levels of interference are kept to a minimum.

The NEWLEOs are confident that they can meet these interference conditions and coexistence criteria and on this basis should be allowed to share spectrum presently used exclusively by these incumbent operators. If this can be made to work, there are substantial social, economic, and possibly political gains that could be achieved.

From a technical perspective, there are compelling reasons for the industry to move towards a mixed constellation model in which users are serviced from a combination of LEO, MEO, and GSO satellites providing always directly overhead (as opposed to nearly always nearly overhead) connectivity.

There is a persuasive argument that 5G terrestrial services could be added to this connectivity offer with all entities coordinated to allow uplink and downlink services and terrestrial services to be mutually complementary. However, there are a legion of regulatory and competition policy and national security and sovereign nation issues and concerns that need to be resolved before this can become a practical reality. In the following chapters, we explore the arguments for and against this mixed constellation model and the related regulatory and commercial implications.

References

[1] http://www.itu.int/en/ITU-R/study-groups/workshops/RSG7-ITU-WMO-RSM-17/Documents/World%20Radiocommunication%20Conference%202019%20and%202023%20issues_V2.pdf.

[2] http://www.itu.int/en/ITU-R/conferences/wrc/2015/Pages/default.aspx.

[3] http://about.dish.com/blog/policy/mvdds-5g-coalition-files-comments-fcc.

[4] http://newsroom.sprint.com/softbank-to-acquire-70-stake-in-sprint.htm.

[5] https://www.orbcomm.com/.

[6] https://www.iridium.com/network/iridiumnext.

[7] https://www.globalstar.com/en/.

[8] https://apps.fcc.gov/edocs_public/attachmatch/FCC-16-89A1.pdf.

[9] http://www.qorvo.com/newsroom/news/2017/qorvo-80211ax-portfolio-provides-broader-faster-lower-cost-wi-fi.

[10] www.erodocdb.dk/Docs/doc98/official/pdf/ECCRep166.pdf.

[11] https://www.itu.int/dms_pub/itu-r/opb/rep/R-REP-SM.2093-2-2015-PDF-E.pdf.

[12] https://www.fcc.gov/document/oneweb-processing-round-initiated.

[13] http://spacenews.com/oneweb-weighing-2000-more-satellites/.

[14] https://www.fcc.gov/document/oneweb-processing-round-initiated.

[15] http://spacenews.com/oneweb-weighing-2000-more-satellites/.

[16] https://en.wikipedia.org/wiki/Reaction_wheel.

[17] http://bluecanyontech.com/reaction-wheels/.

[18] http://www.investopedia.com/articles/economics/09/lehman-brothers-collapse.asp.

[19] https://www.itu.int/newsarchive/press/WRC97/SkyBridge.html.

[20] http://spacenews.com/dish-network-battles-oneweb-and-spacex-for-ku-band-spectrum-rights/.

[21] http://www.fiercewireless.com/tech/dish-led-mvdds-coalition-urges-fcc-to-act-12-2-12-7-ghz-band-for-5g-asap.

[22] https://www.wirelessweek.com/news/2017/06/verizon-locks-early-federal-approval-acquire-nextlinks-28-ghz-spectrum.

[23] https://wireless.fcc.gov/services/index.htm?job=service_home&id=lmds.

[24] http://www.theiaspace.com/.

[25] https://www.appleworld.today/blog/2017/3/11/apple-may-be-funding-boeing-internet-satellite-development.

3

Link Budgets and Latency

3.1 Latency and 5G Standards

Part of the purpose of writing this book is to provide engineers and product planners in the satellite industry with technical details on the 5G standards process, including the New Radio work on the physical layer and the related relevance to next-generation satellite service development. Conversely, I want to provide the 5G community with visibility to the performance potential of satellites for many 5G use cases including counterintuitively some latency critical use cases.

In this chapter, I look specifically at the parts of the 5G standards process that have an impact on latency across the four designated 5G application domains, enhanced mobile broadband (eMBB), low mobility large cell (LMLC), ultrareliable low-latency communications (URLLC), and massive machine-type communications (MMTC). You might think that the latency story would only be relevant for URLLC, but in practice URLLC services can be delivered as a preemptive payload within an eMBB channel, low mobility large cells need to consider round-trip flight time from the base station to the user and/or IoT device. The users and/or devices could be moving at 1,000 km/hr (aircraft) or 500 km/hr (trains).

The most extreme latency requirement in the 5G use cases (covered in more detail in Chapters 9 and 12) is for MMTC IoT connectivity in "the factory of the future" with a minimum value of 100 μs.

In a 5G network it is also important to differentiate between user plane latency and control plane latency. User plane latency is the contribution of the radio network to the time from when the source sends a packet to when the destination receives it (in milliseconds). It is defined as the one-way time it takes to deliver an application layer packet/message from the radio protocol layer 2/3 service data unit (SDU) ingress point to the radio protocol layer 2/3 SDU egress point of the radio interface in either uplink or downlink in the network for a given service in unloaded conditions, assuming that the mobile station is in the active state. The minimum requirements for user plane latency are 4 ms for eMBB and 1 ms for URLLC assuming unloaded conditions (i.e., a single user) for small IP packets (e.g., 0 byte payload + IP header), for both downlink and uplink.

Control plane latency refers to the transition time from a battery-efficient state, for example, an idle or deep sleep state to the start of continuous data transfer, in effect, the time between being asleep and active. The minimum requirement for control plane latency is 20 ms, although there are arguments for reducing this to 10 ms or less. The purpose is to reduce the power drain in battery-driven devices and energy cost and energy consumption in the network. In IoT applications, there can also be a defined time period between wake-up events. For example, if a life of 10 years is required from a button cell battery, a device might only wake up every few hours at a defined moment or for a LEO satellite, every 110 minutes or so as the satellite flies overhead.

Control plane latency also determines the length of time it takes the network to respond to changes in loading condition. So, for example, if offered traffic is bursty, the traffic offered both at the radio layer and network can vary dramatically and rapidly. In an ideal world, radio and network bandwidth would be provisioned to accommodate the most extreme loading conditions, but this would mean the radio layer and network would be underutilized for most of the time. In practice, traffic is buffered, and this introduces delay and delay variability. If there is insufficient buffer bandwidth at any point, then packets will be lost or discarded and will need to be retransmitted. In effect, an IP network is bandwidth-efficient but not inherently deterministic. As soon as we set out to impose determinism on the network, for example, by giving latency-sensitive traffic priority, there will be an associated bandwidth and energy cost that includes additional control plane overhead. Most of us experience the impact of high contention ratios over the internet on a daily and hourly basis so it is no surprise that accessing the internet over a mobile broadband network will have similar performance constraints. All that is different is that the bandwidth limits of the physical layer are determined by the amount of available spectrum and network bandwidth limitations rather than cable, copper, and fiber contention ratios.

Additionally, the 5G standards support a number of dual connectivity user cases that could either help or hinder the delivery of deterministic end-to-end services with defined and closely managed latency parameters.

3.2 Other Factors Influencing Latency

We also need to consider the interrelationship of propagation models, link budgets, device performance, and latency. There are substantial scatter and absorption losses that need to be accommodated in centimeter-band and millimeter-band 5G terrestrial networks that need to be characterized in propagation models and channel models. The propagation models determine the link budget and the link budget determines range and throughput and channel coding overhead, often described in the internet world as goodput (the ratio of user bits to coding and control plane bits). However, link budgets assume that devices meet a conformance standard, for example, receive sensitivity power output and resilience to unwanted signal energy (dynamic range and ability to manage interference). Theoretically, all devices meet their conformance specification, but this is verified by measuring devices directly at the output port of the antenna. In practice, if the devices are tested in an anechoic chamber, an expensive and time-consuming process, they can be shown to perform significantly below the conformance specification, sometimes of the order of 10 dB or more. Conformance specifications may also be relaxed over time, for example, if passbands are made wider or multiple technologies and bands need to be supported in small handheld devices.

Conversely, it is possible that devices work better than their conformance specification. An example would be GSM phones, which though the 1990s generally gained about 1 dB per year of sensitivity, with phones at the end of the 1990s commonly measuring about 7 dB above the conformance specification (−102 dBm). This was a consequence of market scale that allowed tighter tolerances to be imposed on RF component supply chains. Sensitivity then steadily worsened as new bands and new technologies (3G and 4G) needed to be supported. Finally, devices and the component used in devices often fall far short of the performance claimed in the specification sheets because they have been measured in ideal laboratory conditions. Unsurprisingly, the result is that devices and components work less well than expected in the real world in terms of their sensitivity, selectivity, stability, and output power.

The important point to grasp here is that scale helps minimize these implementation issues. More design effort can be applied and supply chains can be bullied to improve raw device performance and the batch-to-batch and device-to-device variability of that performance.

Last but not least, the mechanisms for minimizing interference can be influenced by a wide range of internal and external factors. In mobile broadband systems, for example, including 5G, interference is managed in the frequency and time domain. A TDD network is particularly dependent on maintaining time offsets between interfering devices and accommodating differential delay introduced by flight distance from the device to the base station and multipath.

To be efficient, TDD networks coexisting in the same passband should be clocked together with cosited base stations. This becomes harder to manage as cell size and round-trip time increases and is the reason why all satellite networks separate users and channels in the frequency domain rather than the time domain. However, there are also issues with higher bit rate high user/device density networks. The latest 802.11ax standard, for example, introduces FDD into the physical layer as an additional mechanism for managing localized user-to-user interference but also to accommodate a high density of access points (see Chapter 10 for more details).

TDD timing can be relaxed by increasing the length of the time-domain guard band either side of a transmitted packet of user data but this absorbs radio network time-domain capacity and therefore has an associated cost. Conversely, reducing the time-domain guard band requires a more closely toleranced time reference, which will have an associated cost.

Finally, performance requirements such as operating temperature range can have an indirect impact on latency. Many industrial applications, for example, are required to work over an extended temperature range, which can be as extreme as –40°C to +125°C. This places stress on many of the components in the front end of a device including power amplifiers, low noise amplifiers, filters, and oscillators. Essentially, noise increases with temperature but in the other direction, many components do not perform well when it gets cold, batteries being a significant example.

Satellite engineers must manage far larger temperature gradients and other pesky issues such as radiation damage and the occasional collision but satellites, as I shall show, are a critical part of the end-to-end latency story, both in terms of the user experience and routing and backhaul efficiency.

3.3 Latency, Distance, and Time

5G and satellite operators have significant ambitions to develop vertical markets where latency is a critical parameter which needs to be managed and controlled. It is important to stress that this is only achievable over short distances. In 1 μs, light and radio waves in free space will have traveled 300m so basic physics is going to prevent the delivery of 100-μs latency over more than 30 km and this is before you consider the slower speed of light (and radio) in fiber and routing flexibility.

Table 3.1 shows the time and distance relationship of radio and light waves in free space.

To put this in to a geographic perspective, Singapore is 50 km from east to west and a radio or optical signal will take 166 μs to go from one end of this

Table 3.1
Time and Distance for Radio and Light Waves

Time	Distance	
1 s	300,000 km	186,000 mi
1 ms	300 km	186 mi
1 μs	300m	1,000 ft
1 ns	30 cm	1 ft

high-tech island to the other (Figure 3.1). Malaysia from coast to coast will take 1 ms.

Australia from the east coast to west coast is 4,000 km so that is a coast-to-coast travel time of just over 13 ms. Africa north to south is 8,000 km so a top to bottom time of 26 ms assuming direct routing.

End-to-end latency also depends on the efficient distribution of an accurate time reference across a network, particularly if devices are moving in and out of sleep mode to reduce power drain but also poor clocking will increase the likelihood of a loss of synchronization. The 3GPP vertical market use cases [1] identify a need for more accurate centralized and localized time coordination to support safety-critical automotive transport systems, energy grid applications, e-health and m-health, and factory of the future applications including a requirement to support end-to-end latency of less than 5 ms. Requirements for a 5G network described in the IMT2020 Vision September 2015 include a latency of 1 ms.

Given that light and RF waves traveling in a straight line take 1 ms to travel 186 miles, by the time transmission loss and routing is added in, 5 ms, let alone 1 ms, is an ambitious target over anything other than short distances. However, latency and link budgets are closely coupled.

Safety-critical vertical markets require a 1 in 10^5 packet loss threshold. A packet loss threshold is a combination of the packet loss and the end-to-end

Figure 3.1 Singapore at light speed.

latency constraint. Packet loss rates can be reduced by resending packets but this introduces delay and delay variability. A 1 in 10^5 packet loss threshold might seem a modest target given that fiber is typically specified at 1 in 10^{12} but cellular networks have typically been designed for 1 in 10^3 for legacy voice. Moving from 1 in 10^3 to 1 in 10^6 requires an extra 3 dB of link budget and more closely managed core and edge timing. Every decibel of additional link budget translates into a 14% increase in network density. Reducing the packet loss threshold therefore has a direct impact on capital and operational cost.

3.4 Other Network Overheads and the OSI Model

There is a saying, probably wrongly attributed to Albert Einstein, that the only reason for time is so that everything does not happen at once, although at least Einstein understood the significance of this on a cosmic scale. The need to make sure things do not happen at once is an important aspect of interference management and integration, particularly in TDD networks cosharing the same spectrum, but also FDD where half-duplex is used (frequency-domain and time-domain separation between users). It is also crucial for handover, for channel aggregated multiplexing, and for intercell interference coordination. Work is ongoing on how to time coordinate 5G with LTE Advanced and LTE Pro cosharing the same passband. This will be important both for initial non-stand-alone implementations of 5G coupled to the LTE control plane and later stand-alone implementations, implying a need for control plane coordination. The same time coordination principles could beneficially be repurposed for 5G and satellite interference coordination.

Satellites are often regarded as introducing long latency, but this is an oversimplification. End-to-end delay over a satellite network, particularly a LEO network with inter satellite switching can be quicker than terrestrial in certain conditions. Crucially, there is also the second-order effect of latency variability also sometimes described as jitter.

This can be a bigger problem. A known delay can often be accommodated relatively easily but variable delay can be trickier to manage and can unsettle upper-layer processes including authentication and end-to-end security protocols.

This brings us to the protocol stack and the problem of upper-layer error control. In the late 1970s, before cellular, it was recognized within ISO (the International Organization for Standardization) and the International Telegraph and Telephone Consultative Committee, or CCITT (the abbreviation is from the French version of the name) that there should be a unified standard for describing networking models. It took a while but in 1984 a unified reference model known as the Open Systems Interconnection (OSI) Reference Model (see Table 3.2) was published.

Table 3.2
The OSI Model

OSI Model: Software/Hardware Distribution and Scheduling Response Times				
Layer 7	Application	Windows, Android, Apple	Software	Minutes
Layer 6	Presentation	HTML/XML		
Layer 5	Session	Reservation Protocols		
Layer 4	Transport	TCP prioritization		Seconds
Layer 3	Network	IP address protocols		
Layer 2	Data Link	ATM/Ethernet/MAC		
Layer 1	Physical	Fiber, cable, copper, wireless		Milliseconds

This can still today be universally applied to both guided (fiber, cable, and copper) and unguided (RF and free space optical) physical layers and is still a convenient and effective way of describing the impact of physical layer (Layer 1) impairments on upper-layer performance.

Note that we have added in an arbitrary partitioning between hardware and software with hardware still dominant at the physical layer (the low-cost, power-efficient, software-defined radio is still just around the corner) with the upper layers increasingly implemented as software. As always, it is a trade-off between (software) flexibility and (hardware) performance.

From the point of view of 5G and satellite, the point to make is that any physical layer impairments have a multiple cumulative effect at the upper layers of the protocol stack. A simple example would be automatic repeat requests where the error rate at layer 1 triggers send again requests. In LTE, these repeat requests can typically introduce up to 8 ms of delay, a combination of delay, and delay variability (the delay is an unknown variable). A few automatic repeat requests will trigger upper-layer TCP-IP repeat requests and the result will be reduced throughput, a capacity cost, and additional and unnecessary power drain.

3.5 A Brief History of Time in Mobile Broadband Networks and the Impact on Latency

It seems like ancient history now, but when GSM was introduced in the early 1990s, it was based on a 20-ms frame rate (nicely matched to the voice syllabic rate) supporting a 13-Kb voice codec with 3 Kbps of coding to occupy a 16-Kbps channel multiplexed up to the Integrated Services Digital Network (ISDN) 144-kbps channel rate. 3G introduced the 10-ms time base used in asynchronous transfer mode (ATM) networks. The logic was that 3G networks would need to manage much higher amounts of asynchronous bursty traffic and handle different traffic types and traffic priorities in the same channel multiplex.

4G retains the same 10-ms time base but introduces subframes, two half-frames, each half-frame split into five 1-ms frames and LTE Advanced introduces 1 ms as a time base and 5G reduces this to 0.1 ms based on the concept of a mini slot. The theoretical benefit is tighter control over layer 1 latency, multiplexing efficiency and power efficiency; however, the combination of higher data rates and higher level of time resolution requires a more tightly managed and more accurate and stable time base.

Legacy cellular networks such as GSM have relatively straightforward timing and synchronization requirements with frequency synchronization provided via asynchronous Ethernet backhaul using the IEEE 1588 Precision Time Protocol and or synchronous Ethernet (Sync E).

Distributed timing using Sync E results in frequency synchronization with an accuracy of 50 parts per billion at the air interface, which, in turn, requires 16 ppb at the base station interface to the backhaul network. The introduction of CDMA in the United States introduced an additional need for phase synchronization. This is implemented by using GPS as a frequency and phase reference to an accuracy of between 3 and 10 microseconds depending on the cell radius.

In common with CDMA, LTE TDD and LTE Advanced networks also require phase and time synchronization. In frequency-synchronized networks, pulse transitions happen at the same rate but not at the same time. They can and probably will have a phase offset. In phase synchronized networks, the leading edge of the pulses occur at an identical moment. In phase and time-synchronized networks, the leading edge of the pulses occurs at the same time as the phase transition.

The time and phase reference in LTE TDD and LTE Advanced has to be traceable back to Coordinated Universal Time (UTC) and requires a phase accuracy of 1.5 μs for cell radii of up to 3 km and 5 μs for cell radii over 3 km. This is defined by the ITU standard ITU-T G.8272 and needs to compensate for variable delay introduced by router hardware and routing flexibility. The base unit of UTC is the SI (International System of Units) second. The Si second is defined by a caesium fountain atomic clock.

If you ask a timing expert how accurate a time reference is needed for any given application, the answer will always be "it all depends on…." One of the dependencies is the time over which the timing accuracy needs to be maintained. For example, a time-stamping requirement for a financial transaction or automated computer trading system of less than 1-ms drift compared to UTC can be maintained for 3 hours independently of GPS using a standard temperature controlled oscillator. Maintaining the same specification over 3 weeks requires a high-specification rubidium source [2].

Maintaining <1 μs of accuracy relative to UTC, needed, for example, for high-frequency trading or smart grid or LTE Advanced mobile networks using a temperature-compensated crystal oscillator (TCXO) would support 3 minutes of holdover. Three hours of holdover would need a highly specified oven-controlled oscillator (OCXO) or low-specification rubidium source.

Legacy networks have typically been deployed using a master clock frequency specified to the G.811 ITU standard [3] developed to prevent slips in international switch buffers, primarily for speech traffic, but also used as the master clock for systems such as synchronous digital hierarchy (SDH). This has been supplemented by ITU-T G 8272 for time, phase, and frequency in packet networks and other recommendations in the G.827x series to compensate for variable delay introduced by switch and router hardware and routing flexibility.

Digital networks since plesiochronous digital hierachy (PDH) systems in the 1980s through to SDH and synchronous optical network (SONET) and today's optical networks have required synchronization. These guided media protocols are inherently suitable for synchronization distribution due to the bit-by-bit deterministic way in which they transport data.

The transition to packet networks and Ethernet for backhaul in parallel with the need to maintain legacy TDM networks has meant that synchronization has to be maintained across nondeterministic packet networks.

A common method of achieving this is by using the Precision Time Protocol (PTP) based on a continuous exchange of time-stamped packets, which ensures that the grand master clock reference maintains the alignment of boundary and slave clocks. A parallel protocol, the Network Time Protocol (NTP), is used to synchronize computer clocks over a network.

These protocols can be compromised by frame delay (latency), frame delay variation (packet jitter), and frame loss. PTP operates in a similar manner to NTP, but at higher packet rates and generally at the Ethernet Layer rather than the IP layer. This allows PTP to achieve higher levels of accuracy than the 1-ms level generally quoted for NTP systems [4].

Inconveniently, packet delay in the network is often asymmetric, different between master to slave and slave to master. This complicates the phase synchronization process because the offset computed by the slave will be wrong by the sum of the difference between the two paths.

For example, a computer or server exchanging time stamps every second between a slave and master with 50-ns accuracy could be transitioning through a switch or router introducing asymmetric path delay (packet delay variation) of the order of tens of microseconds.

The computer or server will be running an operating system coupled to a quartz oscillator, which can add microseconds of error per day and there will be an additional difference of several microseconds depending on whether the server is loaded (with the fan running) or unloaded. The filling and emptying of traffic buffers cause additional asymmetric delay variation.

The impact of this is that the core network reference has to be at least an order of magnitude more accurate than the boundary clock reference, for example, 1ms at the edge will need 100 ns at the core. This level of accuracy is also needed to provide back up when GPS is unavailable.

There seems to be an emerging consensus within the 5G standards community that there will need to be a reference time accuracy at the network edge of the order of 300 to 500 ns, which implies 30 ns at the core though it is hard to see how useful this will be if the other causes of end-to-end delay cannot be measured and managed and could potentially result in unexpected edge timing and synchronization costs.

This also implies a need to qualify the timing needs of network function virtualization (NFV), assumed as one of the prime mechanisms for reducing delivery cost in 5G networks; a badly timed virtual network will by implication be a badly behaved virtual network. Packet timing protocols work adequately well over Layer 2 (the data link layer) but not Layer 3 (the network layer) and expensive workarounds may be needed that will negate the promised cost benefits.

The default answer is to use GPS with the comfort and assurance that GPS is becoming more accurate and resilient to jamming with the addition of the L2 and L5 frequencies, launch of the Galileo and Beidou constellations, and enhanced upgraded Glonass but getting GNSS signals into buildings can be hard and expensive. Lightning strikes or high winds can take out external antennas and satellite signals are subject to space weather effects. A wired alternative therefore continues to be a desirable backup. Some countries are investing in additional time reference systems that can act as an additional backup to GPS. The Quazi Zenith constellation being implemented in Japan is one example [5].

Generally, it can be stated that as bit rates increase and the number of users and access points increase, and as more networks are locked together in the time domain with time-domain mechanisms used to manage interference, it will become increasingly important to maintain and distribute an accurate clock reference.

3.6 The Cost of Accuracy

A possible longer-term alternative is to have very accurate clocks distributed through the network and in edge devices. There are emerging low-cost atomic clock options developed originally by DARPA to provide accurate dead reckoning for missiles flying when their GPS reference is jammed. The devices are known as Chip Scale Atomic Clocks. The principle of miniature atomic clocks is based on a technique known as coherent population trapping using a compact sealed vacuum cell of a few cubic millimeters, which contains an alkali vapor which is illuminated by a high frequency modulated laser beam. A device available today produced by Symmetricom [6] uses caesium 133 and a buffer gas in the resonance cell. The vapor is illuminated with a semiconductor laser modulated at a frequency close to the natural oscillation frequency of the caesium atoms, about 9.192 GHz. As the caesium atoms start to oscillate, they absorb less light and the photons transmitted through the cell are used to determine when the modulation frequency of the laser beam coincides with the resonant frequency of the atoms. It is effectively an atomic phase lock loop. The Symmetricom clock weighs 35 grams and draws 115 mW of power and measures 4 by 3.5 by 1.1 cm. It is accurate to within less than half a microsecond a day and can work across a −10°C to +70°C operating range.

This makes it useful for a whole range of applications including backpack military radios, military GPS receivers, unmanned aerial vehicles, backpack IED jammers, and marine geophysical sensors (GPS does not work under the sea). At around $1,500, it has not yet achieved consumer price levels but as prices fall and accuracy improves these miniature clocks will become useful in 5G mobile broadband and telecommunications timing and positioning systems. As with all electrical equipment, these devices will be subject to electrical failure.

Improving the accuracy of grand master clocks is also both desirable and necessary for 5G but has cost implications. An optimized caesium clock costs around $100,000 but caesium depletion means that the caesium tube will need replacing somewhere between every 5 and 10 years at a cost of $30,000.

Strontium-based atomic clocks are being suggested as an alternative [7] as are optical clocks [8] but caesium and rubidium-based devices remain as the default sources of accurate time in present and future networks for at least the next few years.

However, there are other performance parameters including start-up time that are critical to radio network applications including broadcasting, satellite, and terrestrial mobile broadband, which introduce additional synchronization cost. The better clocks typically have much longer start-up and stabilization times.

Resilience is also a cost and generally dependent on supplying multiple clock sources. The repurposing and recommissioning of legacy Loran very low-frequency (VLF) transmitters is being studied and tested as a cost-effective way to provide UTC traceable time to applications in GNSS-denied environments. Initial test program results suggest this could yield UTC traceable results with an accuracy of better than 100 ns, a quality comparable to GPS but with better indoor penetration [9]. Supplementary system innovations such as e-Loran are therefore useful as potential additional time sources.

3.7 Time, Latency, and Network Function Virtualization

We have said that the transition from 4G to 5G implies a need for higher data rates, lower end-to-end latency, better resiliency, lower packet loss thresholds, and low packet delay variability. These together with advanced interference management techniques in the radio layer imply at least an order of magnitude improvement in time accuracy both at the core and in boundary clock devices.

This improvement will also be needed to support network function virtualization (NFV). In particular, the promised cost-efficiency gains of NFV may be at least partially offset by additional synchronization costs. At the very least, it is to be expected that synchronization costs are likely to increase as a percentage of network deployment costs as we move from 4G to 5G networks.

Clock quality is equally critical to all guided media including next-generation cable (DOCSIS 3), copper (G.fast), and fiber (GPON) as is the time-domain integration of the radio access layer with copper, cable, and fiber backhaul. It is plausible to claim and probably possible to prove that clock quality value, the difference between the cost of improving clock quality versus the additional realized value at network and device level increases as bit rate increases. Traffic per watt efficiency will also increase although it is hard to find a costed analysis of this.

3.8 New Radio Specification and Related Latency Issues

The sheer complexity of the 5G standards process can make the analysis of a single aspect, in this case latency, quite difficult, but let us apply our best effort to the process.

At the time of this writing, the freeze date for Release 15 was set at September 2018 with the release specifying eMBB and some parts of URLLC. However, there is a subset of Release 15 which is scheduled for Freeze 6 months earlier known as nonstand-alone where 5G is coupled closely to the time base and frame structure and control plane topology of LTE. Nonstand-alone includes changes to the core network designed to make it more flexible when managing multiple traffic multiplexes with many different latency and throughput requirements.

The specification includes quality of service (QoS) and policy frameworks, both of which will have a direct impact on the latency delivered to individual users and devices, network sharing (which theoretically at least should improve multiplexing efficiency), and trust hierarchies. Incidentally, authentication protocols can have a major impact on latency as they can introduce milliseconds of delay at a local level and seconds of delay at network level. We have also made the point earlier that latency and delay variability can compromise the authentication process, more or less ensuring that a session never gets started.

At the physical layer, the important time domain difference at frame level is the introduction of the mini slot with the specific objective of making present and future URLLC requirements easier to meet. For example, a mini slot can be used for URLLC preemption in an eMBB shared channel. The mini slot is also intended to help operation in unlicensed bands by allowing transmission to start directly after a successful listen-before-talk procedure without waiting for the slot boundary.

Mini slots consist of two or more symbols; the first symbol includes uplink or downlink control information. The minimum scheduling intervals scale as the subcarrier subspacing increases from 1 ms with 15-kHz subcarrier spacing to below 0.1 ms for 240-kHz and 480-kHz subcarrier spacing. An extended cyclic prefix option for larger cells or less closely managed timing has a scheduled interval of 0.25 ms.

The TDD frame structure is slightly different and divides 20-ms frames into two 10-ms subframes. Note that one advantage of TDD is that the uplink and downlink are reciprocal (on the same center frequency), which makes channel sounding simpler, which, in turn, makes it easier to beam form.

Note that the control channel will generally be more heavily coded than a data channel. The modulation and coding used on the data channel will adapt to changing channel conditions to minimize the triggering of physical layer and higher layer send-again messages. The accuracy and speed with which a network is able to make channel quality measurements are therefore other factors with an impact on end-to-end latency.

Much work continues to be invested in optimizing channel coding to improve error detection and correction. The end objective is to minimize residual bit error rate as the signal to noise and carrier to interference conditions become more adverse. Generally, all schemes combine some block coding with parity checks and convolutional coding. Block coding introduces some delay and convolutional coding (coding with memory) introduces some delay variability.

The reason the parity checks are short is to minimize the impact on physical layer latency, techniques such as chase combining [10] are used to limit the number of clock cycles and time needed to error detect and correct a wrongly decoded symbol. Understanding polar codes really requires a higher degree in mathematics; see [11] for more information.

In an ideal world, signal to noise levels and carrier to interference would be maintained at a level that would minimize bit errors on the radio path. This is relatively easy to achieve in guided media such as fiber, cable, or copper because the impairments are predictable and stable and can therefore be managed and mitigated. This is why a fiber physical layer can be held at a 1 in 10^{12} bit error rate. Unguided wireless links particularly in mobile networks are harder to manage because impairments such as multipath phase cancellation change rapidly and unpredictably. Channel coding in mobile networks allows user bit error rates to be minimized in poor signal to noise or carrier to interference conditions, but there is a bandwidth cost (the addition of extra error correction bits) and latency cost (send again instructions when error rates exceed a certain threshold).

3.9 In-Band Backhaul

Having temporarily exhausted the topic of 5G physical layer frame structures and channel coding (we come back to the topic in later chapters), it is time to take a look at terrestrial backhaul and its impact on end-to-end latency.

Backhaul can either be via microwave link or fixed copper or fiber or free-space optical or, if none of these options is cost-effective, then via satellite. The difference in 5G is that it makes a lot of sense to use the same passband to support users and backhaul. This is known generically as in-band backhaul or self-backhauling and has many advantages including the opportunity to reuse RF hardware and baseband processing. Figure 3.2 illustrates the concept.

Figure 3.3 shows how self-backhauling is integrated into a 5G network

Self-back hauling avoids the need to demodulate and modulate and channel code traffic as it moves from the radio layer into the backhaul network. Note that backhaul in existing networks is often at 28 GHz or 39 GHz so the RF hardware already exists. Given that satellites are also using this spectrum, then there is an obvious opportunity to intensively reuse the spectrum with the caveat that cross-system interference has to be managed. It is an ongoing

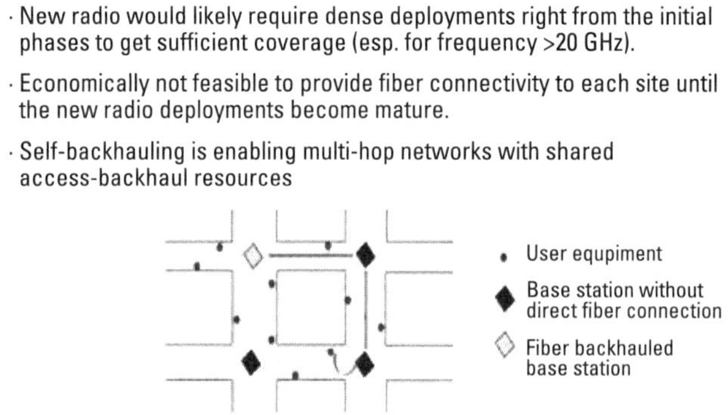

Figure 3.2 Self-backhauling for millimeter-wave cellular. (Thanks to Nokia Networks.)

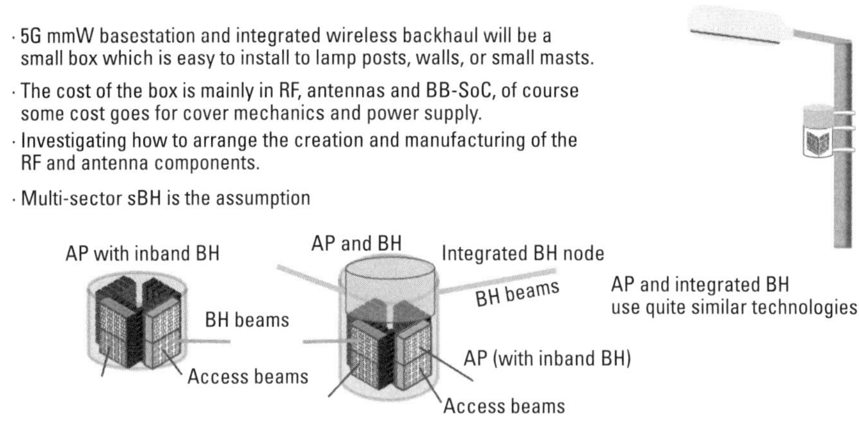

Figure 3.3 Basic network building blocks. (Thanks to Nokia Networks.)

narrative in this book that this process is critically dependent on achieving effective spatial separation between potentially interfering systems and for users of those systems to be confident that these techniques can be made to work across multiple systems administered and managed by multiple operators.

In a dense or ultradense network, it is unlikely that fiber would be economic at least for initial rollout. The self-backhauling is used to get to the nearest fiber end point. If well designed and implemented, this topology should not add materially to the overall latency budget.

Overall, it can be seen that much effort is being invested in standardizing the 5G physical layer and the supporting network topology in order to meet assumed 5G end user end-to-end latency and throughput requirements. This work and the calculated throughput and latency metrics achievable provide a benchmark against which GSO and NEWLEO operators will need to measure their constellation performance.

This brings us logically to a comparison of 5G terrestrial channel modeling and propagation and satellite channel modeling and propagation. Superficially, it might be considered that if a terrestrial system and satellite system are implemented in the same passband, the two systems will have similar propagation constraints. If implemented with similar channel bandwidth and channel spacing then it might be thought that would have similar channel characteristics.

However, the terrestrial user to 5G base station path is different to the terrestrial user to space path. Additionally, different satellite constellation topologies will have different propagation properties, but as we shall see, the angle of arrival and angle of departure of all systems, both terrestrial and space-based, ultimately determines throughput, end-to-end latency, and carrier-to-interference coexistence.

3.10 5G and Satellite Channel Models

3.10.1 3GPP TR 38.901

The starting point in this discussion is the 5G channel model. These are described in the 3GPP document TR 38.901.

3.10.2 Line of Sight and Nonline of Sight

If you wanted a one-line summary of the issues addressed in this document, then it would be reasonable to say that 5G in the centimeter and millimeter band works better when line of sight. Unfortunately, this is a rare occurrence in urban and rural topographies, both for mobile and fixed users and even line of site from a low elevation, for example, a base station a few meters from the ground can be subject to substantial nonadditive ground reflection. This is the single most important advantage that high-count LEO constellations in the K-bands or V-band and W-band/E-band have over terrestrial 5G because, as we have stated earlier, these constellations will be nearly always nearly overhead. If 10,000 LEO satellites are launched into space, then there will be a satellite directly overhead more or less all the time more or less anywhere in the world.

This not only minimizes the Earth-to-space latency budget but minimizes the path through the atmosphere. The signal will also be minimally affected by surface scatter or ground reflection. Frustratingly, it is presently hard to quantify exactly how much power from a terrestrial RF transmitter gets lost on its way to its local destination.

3.10.3 Existing Models

Contemporary cellular networks at ultrahigh frequency (UHF) or L-band and C-band are designed using well-established propagation models with physical layer RF and baseband parameters determined by a range of user defined pedestrian typically urban (TU3), vehicular urban (TU50), and rural channel models (RA250). These work adequately well up to 4 GHz but become progressively less accurate at higher frequencies/shorter wavelengths.

Discussions around suitable channel models for the centimeter and millimeter bands focus on the relative merits of the alpha beta gamma (ABG) model using a floating constant referenced to known and measured data sets, the close-in (CI) model referenced against a path distance of 1m and a frequency weighted path loss exponent (F), but no large-scale existing data sets are presently available to verify/fine-tune these models. Anecdotally, the observation is made that the ABG model typically underpredicts path loss when near to the transmitter and overpredicts path loss further way. The CI and CIF models are more accurate and computationally simpler. The CI model works better for outdoors and CIF model works better for indoors. Both models have a path loss variable that is continuously coupled to the transmitted power over distance.

Within the EU, research work has been focused on frequencies from 2 GHz to 73 GHz with a path length of between 4m and 1,238m with models for urban microcells (UMi) with antennas at 10m, urban macrocells (UMa) with antennas at 25m, and indoor hot spots (InH).

The measurements and modeling are based on narrow beam 7.8° azimuth half-power beamwidth antennas and wideband 49.4° antennas. While this work will almost certainly yield useful outputs it does not include modeling for larger cells. Neither does it set out to model fractional beamwidth antennas (half-power beamwidth of between 0.5° and 1.5°).

Fortuitously, substantial modeling does exist for point-to-point backhaul, which yields simple but useful path loss estimates for specific frequencies for line-of-sight and nonline-of-sight links in a range of atmospheric conditions (rainfall rates) and taking into account oxygen absorption (peaking at the oxygen resonance frequency at 60 GHz) as shown in Table 3.3. Note how the wavelengths above 30 GHz to 60 GHz (10 mm to 5 mm) are similar to or less than the roughness of many man-made and natural surfaces, hence the high absorption and surface scatter.

As a rough but useful rule of thumb, a 28-GHz receiver needs an additional 30 dB of isotropic gain to see the same amount of power as a 900-MHz receiver. As we shall see in Chapter 6, this is not that hard to achieve. To restate, the main enemy here for 5G terrestrial is surface scatter and absorption and nonline-of-sight losses.

3.10.4 ITU Rain Models and Satellite Fade Calculations

The ITU rain models are well established and based on extensive measurements of fading in a range of weather conditions including monsoon (>150 mm/hr), tropical rain (100 mm/hr), heavy rain (25 mm/hr), light rain (2.5 mm/hr), and drizzle (0.25 mm/hr) with fading ranging from fractions of a decibel to

Table 3.3
Centimeter-Band and Millimeter-Band Propagation Measurements and Modeling Used to Design Wireless Backhaul

Frequency (GHz)	Wavelength	Path Loss (dB)		Rain Attenuation		Oxygen Absorption
		Line of sight	NLOS	5 mm/h		At 200m
28	1.07 cm	1.9	4.6	0.18 dB	0.9 dB	0.04 dB
38	7.89 mm	2.0	3.8	0.26 dB	1.4 dB	0.03 dB
60	5 mm	2.23	4.19	0.44 dB	2.0 dB	3.2 dB
73	4.1 mm	2	2.69	0.6 dB	2.4 dB	0.09 dB

over 10 dB and in exceptional conditions at higher frequencies for many tens of decibels.

These models can be applied accurately to line-of-sight terrestrial microwave links but are less appropriate for satellite modeling due to the need to accommodate a range of elevations.

A low-elevation angle will suffer more rain fade more often due to the longer path length through the atmosphere. Conversely, if a directly overhead link passes through thick thunderclouds, then a lower elevation link will have lower path loss. This is one example when an active antenna will deliver additional gain over a passive antenna which can only look directly upwards.

3.10.5 Oxygen Resonance Lines and the Very High Throughput V-Band Duplex Passbands

For the sake of completeness, Figure 3.4 is the much-reproduced graphic showing the water and oxygen resonance lines.The increase in propagation loss at 60 GHz is used to maximize frequency reuse.

The windows where loss is lower determine the optimum position for V-band very high throughput constellation proposals. For example, the Boeing V-band proposal has a lower duplex passband a 37.5 to 40 GHz and an upper duplex passband at 51.4 to 52.4 GHz.

An additional issue is that channel measurements need to be made across a wide range of channel bandwidths from a few kilohertz to 1 or 2 GHz. The additional cost of wider-band spectral analysis means that there tend to be more narrowband measurements available, which can be misleading.

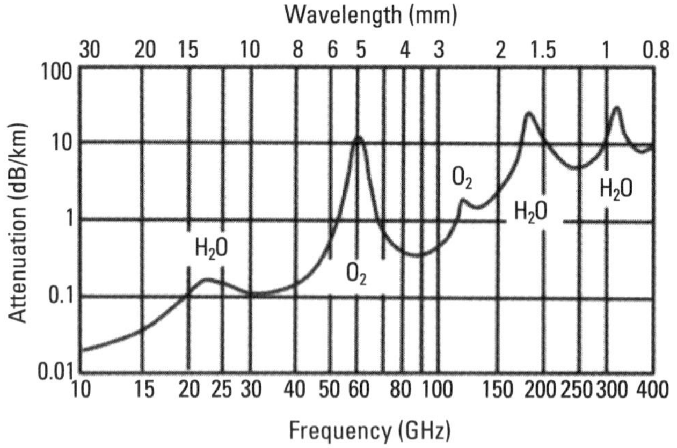

Figure 3.4 Water and oxygen resonance lines. (Thanks to the Rutherford Appleton Laboratory.)

3.10.6 Beyond Line of Sight

To complete the list of channel model variants, we should also include beyond line of sight (BLOS). This applies both to radar systems (over the horizon radar) and two-way communication systems that rely on radio waves following the curvature of the Earth. This phenomenon was discovered by British radar engineers working around 300 MHz in the 1930s and resulted in a correction figure known as K factor, which calculates the propagation effect as being equivalent to increasing the diameter of the Earth by 33%. The bending is proportional to frequency with waves bending less as the frequency increases, although it can also be affected by atmospheric conditions, which also have a progressively larger impact at higher frequencies.

3.11 Satellite Channel Models and Signal Latency

The story so far is that propagation conditions become more variable as frequency increases, propagation losses are higher, but other factors such as surface absorption and scatter and atmospheric conditions also become increasingly important. Networks need to work in the rain.

These factors are a dominant influence in deciding which of the satellite constellation options could be best suited to complement 5G networks. The options include geostationary satellites, medium Earth orbit satellites and low-count and high-count LEO constellations. In all cases, the channel models and propagation characteristics will be determined by the elevation angle as seen from Earth.

Self-evidently, a terrestrial device looking at a GSO satellite over the equator will be subject to increasing propagation loss at higher latitudes. As latitude increases, the path length gets longer and the signal will pass through more of the atmosphere. The coverage pattern from the satellite will be roughly circular over the equator and increasingly elliptical at higher latitudes so the flux density will be lower at higher latitudes. The longer path length will also increase end-to-end delay.

Low-count MEO and LEO constellations will also be serving users on earth at a variety of inclination angles depending on their position in the sky at any given moment. As mentioned several times already, high-count LEO and MEO constellations will be nearly always overhead nearly all the time.

The speed of light in free space is constant bar some gravitational effects, so flight distance and flight time for all satellites in all orbits can be precisely calculated and are determined by orbital altitude and elevation angle.

GSO satellites orbit the Earth at 36,000 km above the equator. Radio waves go at the speed of light, which is 300,000 km per second. For users on the equator communicating with a satellite directly overhead, the total distance, single hop (up and down) is 72,000 km so the time delay is 480 ms for a round trip.

A geostationary satellite is visible from a little less than one-third of the Earth's surface and if you are located at the edge of this area, the satellite appears to be just above the horizon. The distance to the satellite is greater and for Earth stations at the extreme edge of the coverage area, the distance to the satellite is approximately 41,756 km. Communicating with another similarly located site is nearly 84,000 km away, so the end-to-end delay is almost 280 ms (one way). Extra delays occur due to the length of cable extensions at either end and if signals are routed by more than one satellite hop. Significant delay can also be added in routers, switches, and signal processing points along the route.

In a MEO network (using O3b as an example), orbit height is 8,062 km. A typical single hop path involves sloping path lengths of 11,000 km producing a single-hop distance of 22,000 km producing a latency of 73 ms. O3B claim [12] a round-trip latency of better than 150 ms based on a double-hop distance of 11,250 + 11,250 + 11,250 + 11,250 km.

In LEO networks, the propagation delay is smaller still. Iridium's constellation operates at 780 km, Orbcomm is a little higher at 825 km, and Globalstar is at 1,414 km. The propagation delay experienced in a LEO satellite system varies as the satellites change position but will be 4.3 ms per hop for Iridium, 4.5 ms for Orbcomm, and 7.8 ms for Globalstar for bent pipe (with the satellite operating as a repeater rather than a relay [no onboard signal processing]) applications with the satellite directly overhead. These figures should be doubled for round-trip delay.

If the terrestrial end points are not within the coverage of a single satellite (this varies with each system), then the distance will be greater, with intersatellite links via other satellites.

Propagation delay is only one part of the delay budget. Delay and delay variability are also introduced by processing delay, for example, through any router nodes or relay transponders. If these devices are software configurable, then delay variability could be significant. This is sometimes described as serialization delay. The delay through relay transponders is a function of forward error correction and modulation.

Satellite systems over the past 20 years have evolved from initially using Viterbi coding, then Reed Solomon coding, and then turbo codes (codes with memory). As data rates increase, block sizes increase, convolutional coding delivers more coding gain and avoids send again loops that introduce delay variability. Satellite TV provides a noticeable example of propagation and coding delay when compared to terrestrial TV.

The trend in satellite schemes below 2 Mbps is to use smaller forward error correction block sizes. These can reduce a typical 200-ms error correction induced delay to 50 ms. In other words, 300-ms can be eradicated from the round-trip delay by changing the forward error correction scheme.

Finally, TCP/IP uses acknowledgments to determine the amount of bandwidth available before any data is exchanged between two points (the slow start algorithm). This requires three round trips, or six satellites hops, just to get started and if the session goes idle, the whole process must start again. There are various ways round this including TCP/IP fast start and caching and local storage in off-peak periods. These approaches are now standardized with the Space Communication Protocol Standard [13]. There are also a number of WAN optimizer and WAN accelerators from a cross-section of vendors typically implemented at Layer 4 of the OSI protocol stack.

This would suggest that satellites introduce substantial additional delay over and above terrestrial networks, but a satellite constellation provides absolute visibility across the whole end-to-end channel, which means that delay and delay variability can be calculated and compensated. This is harder to achieve in terrestrial networks.

This may seem trivial, but is actually an important point. Often, it is not the actual delay that is the problem but rather the second-order effect of delay variability. It is hard to design challenge and response and authentication or send-again algorithms when these variables vary over time.

A simple example is me asking you if you are called John. I expect an answer within a certain elapsed time. If you fail to answer for half an hour, I would have a nagging doubt that either your name was not John or if it was you did not want to admit it.

In a communications network, authentication challenge and response algorithms have specific expectations of how long each part of the authentication should take. This is a simple but effective way of reducing the chance that the algorithm has been spoofed. A satellite link with a known end-to-end delay and known and calculable delay variability should be more secure than a flexibly routed exchange across multiple terrestrial networks.

3.12 Ongoing Satellite Standards and Related Study Items

In September 2017, a standards group studying the potential touch points between 5G New Radio standards and nonterrestrial networks finally ground into action. The work group has a rather narrow group of companies involved (Thales, Fraunhofer, Dish Networks, and Ligado) (and no obvious participation from the mobile broadband terrestrial standards community), but at least it is a start.

The 3GPP TR 38.811 2017-06 study [14] on New Radio to support nonterrestrial networks covers bent-pipe payloads and regenerative satellite topologies, which are satellite terms for relays and repeaters. A bent pipe takes a signal from Earth, amplifies it, and sends it back again. A regenerative transceiver demodulates and decodes and then modulates and recodes the downlink.

A regenerative transceiver on a satellite therefore introduces additional processing delay though this is immaterial when compared to the round-trip delay. The advantage is that the signal is cleaned up prior to its onward journey. The bent pipe is therefore effectively performing the same function as an LTE repeater. A regenerative transceiver is analogous to an LTE relay. In a bent pipe, one-way propagation delay is the sum of feeder link propagation delay and user link propagation delay. A regenerative payload is essentially similar, but with onboard processing delay added in.

3.13 Propagation Delay and Propagation Loss as a Function of Elevation

The ongoing narrative is that there are significant advantages to having an elevation angle as close as possible to 90° (directly upwards). For example, a geostationary satellite at 35,786 km if viewed at an elevation angle of 10° will have a path distance of 40,586 km with a delay path of 135 ms versus 119 ms on the direct path.

A LEO at 600 km, if viewed at an elevation angle of 10°, will have a path length of 1,932 km with a path delay of 6.44 ms versus the 2 ms of one-way delay on the direct path.

The shorter direct paths reduce atmospheric fading (water vapor and oxygen absorption, as shown above), rain attenuation, cloud attenuation, and scintillation. Note that scintillation, caused by small scale (tens of meters to tens of kilometers) perturbations in the electron density along the signal path, is more prevalent at low and high latitudes [15].

For NGSO satellites, Doppler also needs to be accommodated. A LEO viewed at a lower elevation angle will either be moving towards the observer or away and will be traveling at about 7,000m per second. This produces a differential delay of 40 μs per second.

If satellites were to use OFDM with a cyclic prefix (CP-OFDM), this would be challenging due to the fact that all users need to be received at the base station aligned with each other within the limits of the cyclic prefix, 4.7s for the standard cyclic prefix and 16.7s for the extended cyclic prefix. This could be achieved using timing advance, but the control plane signaling load would be substantial.

However, if LEOs are viewed directly overhead, for example, in a high-count LEO constellation with fast handover between satellites as they move into and out of the cone of visibility, then Doppler will be moderate (similar to terrestrial LTE) and easily accommodated within the standard LTE timing advance protocols. However, it can also be seen that large differential path lengths can be problematic given that a standard-length prefix is equivalent to a path length of 1.2 km and an extended-length prefix is equivalent to 5 km.

The answer is not to think in traditional cell planning terms but to think about beam planning to ensure that all users served by each individual beam can be time-aligned within the standard LTE frame structure and timing requirements.

3.14 The Impact of NEWLEO Progressive Pitch on Latency and Link Budgets

This highlights the importance of modeling the impact of progressive pitch where the elevation angles are altered to minimize interference into ground base gateways. It can be seen that if low elevation angles are required to meet high protection ratios, then there will be a directly associated performance cost. The alternative approach discussed in Chapter 6 is to use antennas on ground stations and user terminals to separate out wanted from unwanted signal energy with LEO, MEO, and GSO satellites all working as closely as possible to their maximum elevation angles.

3.15 Satellites and Subcarrier Spacing

In an ideal world, it would be useful to have the same or at least a similar physical layer for satellites and 5G. 5G as specified in the New Radio specifications effectively uses the same downlink modulation (CP-OFDM) as 4G LTE. In 5G, CP-OFDM is also used on the uplink. CP-OFDM is not particularly power-efficient but is reasonably robust to phase noise and Doppler. Phase noise increases in proportion to carrier frequency, and frequency drift due to local oscillator inaccuracy and Doppler spread also increases with frequency.

It might therefore be reasonable to assume that subcarrier spacing in 5G satellites would and could be similar to 5G terrestrial with higher subcarrier spacing used at higher frequencies to minimize and mitigate inter carrier interference, for example, 15, 30, and 60 kHz at S-band and 60, 120, 240, and 480 kHz at Ku-band, Ka-band, V-band, and E-band.

3.16 Edge Computing, Above-the-Cloud Computing: The Dot.Space Delivery Model

5G vendors promote edge computing as the mechanism for delivering very short latencies, of the order of a millisecond. This is achieved by placing a server in the 5G node B or LTE node B.

The assumption is that search requirements have sufficient predictability to make this approach cost-effective. If user needs can be satisfied from locally cached content, then this reduces traffic loading into the network including backhaul traffic.

Satellite connectivity for obvious reasons (hundreds of kilometers of path length) can never deliver 1-ms latencies; however, these latencies are only achievable if the required content is in the cache and a good line-of-sight link exists between the node B and the user.

The application equivalent in the satellite industry is to put a server in space, essentially above the cloud connectivity (the dot.space proposition). Elsewhere in this book, I reference companies such as Planet.com [16], which collects and adds value to Earth imaging and sensing data. Presently, this data is brought back to Earth, however servers in space are eminently practical as and when they become economic.

3.17 Summary

Many factors determine latency and delay variability in 5G terrestrial and satellite networks including link budgets and path length. Counterintuitively, satellites can have a link budget advantage provided that they are almost directly overhead, for example, for GSO satellites over the equator or for high-count LEO and MEO constellations at any latitude.

Nearly always nearly overhead or ideally always overhead will generally give a good line of sight into most outdoor coverage scenarios and surface absorption and ground reflection will be minimal.

For the same reason, satellites are not effective at providing connectivity to handheld devices in buildings due to the high penetration losses of typical building structures. The only time that satellites would have visibility inside a building is if they were visible at a low inclination angle and terrestrial networks would probably be more effective in these conditions. Satellite operators overcome this constraint by supplying antennas to mount on window ledges or building roofs. VSAT terminals for corporate connectivity, for example, have been used for many years, but the shift is now towards low-cost transceivers with integrated Wi-Fi for local in building coverage. The OneWeb model for low-cost connectivity outdoors can reuse this in building hardware with a Ka-band transceiver coupled to Wi-Fi. This means that Wi-Fi will add delay to the latency budget, although this should not be material in the context of the overall round-trip delay.

Although round-trip delay might seem to eliminate satellites from many 5G latency-critical use cases, there are circumstances in which satellites deliver a performance advantage. Examples already referenced in earlier chapters include Iridium and potentially LeoSat where the faster speed of light and radio in free space (compared to fiber) outweighs the path delay over distances of more than 10,000 km. Iridium also has intersatellite switching (K-band) and LeoSat and SpaceX have proposed to use optical intersatellite switching, all of which provide absolute control of the end-to-end channel, eliminating the second-order effect of delay variability. This can be particularly useful if high-level authentication is needed. This combination of end-to-end control and security is a basic requirement for military and life-critical systems but also for high-value financial services such as high-frequency trading.

Latency is also a function of the loading imposed on a network. This is known as fill factor in the satellite industry but can also be expressed in all wireless and wireline networks as contention ratio, which is a function of the number of users cosharing a radio channel, fiber, cable, or copper resource or router node. The number of users and the burstiness of the aggregated offered traffic from these users define the required delivery and buffer bandwidth of the network. The buffers are provisioned to allow for queuing so that packets do not get lost or discarded when there are insufficient delivery resources at any point in the end-to-end channel. In an ideal world, a network would be dimensioned to avoid queuing. This would also avoid the need for traffic prioritization and preemption, the process of separating out latency-tolerant and latency-sensitive traffic, which absorbs bandwidth and power. However, this would involve dimensioning the network to accommodate worst-case loading, which would be expensive both in terms of capital and operational cost. Latency is therefore partly a product of network topology but also a product of network economics.

Bearing all this in mind, the question to answer is whether satellites can deliver services to Earth at an equivalent or lower cost than terrestrial 5G. The following chapters set out to find the answer to this question.

References

[1] http://www.3gpp.org/news-events/3gpp-news/1839-5g_cc_automation.

[2] http://www.chronos.co.uk/index.php/en/time-timing-phase-monitoring-systems.

[3] https://www.itu.int/rec/T-REC-G.811/en.

[4] http://www.ntp.org/.

[5] http://qzss.go.jp/en/.

[6] https://www.microsemi.com/product-directory/clocks-frequency-references/3824-chip-scale-atomic-clock-csac.

[7] http://www.nist.gov/pml/div689/20150421_strontium_clock.cfm.

[8] https://www.rp-photonics.com/optical_clocks.html.

[9] http://www.chronos.co.uk/index.php/en/delivering-a-national-timescale-using-eloran.

[10] ftp://www.3gpp.org/tsg_ran/WG1_RL1/TSGR1_17/Docs/PDFs/R1-00-1428.pdf.

[11] http://pfister.ee.duke.edu/courses/ecen655/polar.pdf.

[12] https://www.o3bnetworks.com/technology/latency-throughput/.

[13] http://www.scps.org/.

[14] https://portal.3gpp.org/desktopmodules/Specifications/SpecificationDetails.aspx?specificationId=3234.

[15] http://www.swpc.noaa.gov/phenomena/ionospheric-scintillation.

[16] https://www.planet.com/.

4

Launch Technology Innovation

4.1 Introduction

On September 5, 1859, Isambard Kingdom Brunel (1806–1859) was photographed on board the SS *Great Eastern* [1]. It was the last photograph taken before his premature death, worn out by 30 years of building things that no one else had tried before. In 1859, the *Great Eastern* was the largest ship ever built, 700 feet long with a displacement of 22,500 tons, and it was the first ship built entirely of iron. It was potentially capable of carrying 4,000 passengers from the United Kingdom to Australia on a single load of coal.

In practice, the ship had a checkered history and lost money for several owners but finally found its niche laying transatlantic cables from Britain and Europe to America and across the Indian Ocean.

The point of the story is that, rather like commercial scale, physical scale can deliver a potential gain in performance, in this example speed and efficiency and tonnage capacity, but the end application is not always quite what was originally intended. It is the same with rockets. In this chapter, we write about the new generation of superlarge, superpowerful, superfast, superefficient rockets being developed for the first manned mission to Mars.

This may or may not happen but the rockets can also be repurposed to take massive payloads into near-Earth orbit at a cost far below existing launch systems. This chapter takes us through this evolution process and its impact on satellite economics.

4.2 The Old Rocket Men

The principles of orbital mechanics were first explained in detail by Isaac Newton in his 1687 *Principia Mathematica*, imagining a cannon ball being fired horizontally from cannon on a mountain with enough power never to hit the ground, falling to Earth at the same speed as the Earth falls away from it. Free from atmospheric drag, the cannon ball would orbit the Earth forever [2].

4.2.1 Charles C. Clarke and the Role of Science Fiction

In the October 1945 edition of *Wireless World*, the article by the science fiction writer Charles C. Clarke, "Extra-Terrestrial Relays: Can Rocket Stations Give Worldwide Radio Coverage?" popularized the idea of geostationary satellites broadcasting radio and TV programs. The satellites would be launched from a new generation of rockets capable of accelerating their payload past the orbital insertion velocity of 8 km per second (5 miles a second) [3].

4.2.2 Jules Verne and Herr Oberth

Eighty years earlier in 1865, Jules Verne had published *Earth to the Moon*. Inspired by the advances made in ballistics in the American Civil War, Mr. Verne tells the story of the Gun Club, a group of missile experts who decide to build a giant cannon powerful enough to hit the Moon. In the book, Mr. Verne also mentions that this will be done from Florida and that a manned mission would be launched from the cannon with three crew members flying in an aluminum spacecraft. This was exactly 100 years before the Apollo 8 Moon mission launch.

In 1905, an 11-year-old Romanian, Herman Oberth, contracted scarlet fever and was given Jules Verne's book to read to take his mind off feeling ill. Two years later, at the age of 14, Mr. Oberth proposed a recoil rocket that could propel itself by expelling exhaust gases from its base. Oberth moved to Germany and joined a medical unit in World War I, returning to university shortly after to study mathematics and physics and realizing that what rockets needed was multiple stages. "If there is a small rocket on top of a big one, and if the big one is jettisoned and the small one is ignited, then their speeds are added."

4.2.3 Herr Oberth and Herr von Braun

From 1923 to 1929, Oberth worked on his book, *The Rocket into Planetary Space* [4], explaining how rockets could escape the Earth's gravitational pull. He received a patent for his rocket design and launched his first rocket near Berlin on May 7, 1931, and became a mentor to a young Mr. Wernher von Braun. Hermann Oberth died in Nuremburg, West Germany, on December 29, 1989, at the age of 95. Born in 1912, Werner von Braun shared Oberth's appetite for turning science fiction into fact and in 1928 had joined the German Society for Space Travel.

In 1932, Von Braun went to work for the German army to develop liquid fuel rockets and over the next 10 years developed the V2 rocket (Figure 4.1), a liquid-propelled (alcohol and liquid oxygen) missile, 14m long, weighing 12,000 kg, capable of flying at 5,600 km per hour, delivering a 700-kg warhead to a target (Paris then London) up to 300 km away. The motor typically burned for 60 seconds, pushing the rocket to around 2 km per second, rising to an altitude of about 90 km (Figure 4.2).

4.2.4 Robert Goddard and War of the Worlds

On the other side of the Atlantic, Robert H. Goddard [6] (1882–1945) inspired by H.G. Wells' book, *War of the Worlds* (1897), had by 1914 filed two U.S. patents, one for a rocket using liquid fuel and the other for a two-stage

Figure 4.1 The V2 rocket. (Image courtesy of V2 Rocket.com.)

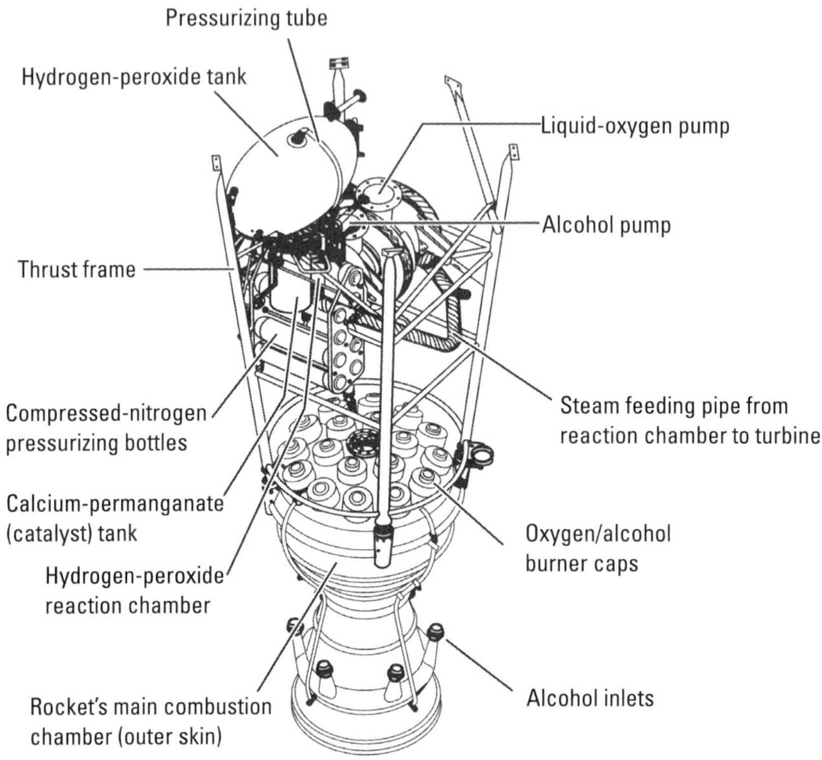

Figure 4.2 The V2 rocket motor [5]. (Image courtesy of V2rocket.com.)

or three-stage rocket using solid fuel. In 1920, the Smithsonian published his paper, "A Method for Reaching Extreme Altitudes," which included a study of the practicality of sending payloads to the Moon. He developed and launched what was probably the world's first liquid-fueled rocket in 1926 in Auburn, Massachusetts. Goddard proved that a rocket would work in a vacuum (Figure 4.3), put the world's first scientific payload on a rocket in 1929, used vanes in a rocket motor for guidance in 1932, developed pumps suitable for rocket fuels, and in 1937 launched a rocket with a motor pivoted on gimbals controlled by a gyro mechanism.

4.3 Red Army Rockets

In parallel, in the early 1920s, Lenin had decided to equip the Red Army [7] with new weapons including solid fuel rockets that could compete with conventional artillery. Stalin from 1924 until his death in 1953 was equally keen on keeping up or ahead of U.S. and German rocket research sponsoring the

Figure 4.3 Robert with a rocket in Roswell, New Mexico, October 1935. (Credit: NASA/Goddard Space Flight Center.)

grandly titled Society for the Advancement of Defense, Aviation, and Chemical Technology in the early 1930s.

The Russian winter proved to be a more effective weapon against Germany in World War II, but the postwar period became the Golden Age or Dark Age of Russian ballistic missile development, depending on your point of view.

The Cold War spurred on a remarkable development effort on both sides of the Atlantic. In Russia, the Soviet rocket designer Sergei Korolev spent from 1954 to 1957 developing the world's first intercontinental ballistic missile, the R7, powerful enough to deliver a nuclear warhead to the United States or launch a spacecraft into orbit, with Sputnik (see Chapter 1) being its first big success. In 1961, a modified R7 launched Yuri Gagarin into space. Yuri was a big pin-up poster hit for teenagers in 1961.

4.4 The German Rocket Legacy

Meanwhile, Herr Braun had expeditiously moved from a stricken postwar Germany to the United States to work for the U.S. Army developing the first generation of Jupiter C, Juno 11, and Saturn 1 launch vehicles with Jupiter C used to put the first U.S. satellite, Explorer 1, into orbit in 1958.

In 1960, President Eisenhower moved rocket development away from the control of the army to the newly established National and Aeronautics and Space Administration (NASA) with a mission to produce a new generation of giant rockets to be known as the Saturn rockets. Herr Braun became the director of the Marshall Space Flight Center [8] and the chief architect of the Saturn V launch vehicle.

4.5 The French and British Legacy

Sixty years on, this research and development behemoth is producing inspiring research on materials and manufacturing innovation, which is literally fueling the new space revolution [9]. Wernher Von Braun died in Alexandria, Virginia, on June 16, 1977. His involvement in the Nazi party always remained controversial, but his contribution to the U.S. space industry is unparalleled.

This is not purely a story of competing super powers. In 1958, Charles De Gaulle, for some the hero of the anti-Nazi resistance movement during World War II, returned to power in France and convinced himself that it would be naïve to depend on the United States for military protection in the era of nuclear-tipped intercontinental ballistic missiles, at which point he committed France to full technological independence from the United States including complex and expensive research programs on nuclear technology and rocketry.

You could debate the longer-term wisdom of this decision, although a probably unexpected outcome is that 60 years later France has an electricity grid fed by a higher percentage of nuclear power than any other country in the world [10]. The guilty secret here is that you can only get a decent supply of weapons-grade plutonium if you have a scale-efficient uranium-processing capability.

Britain was similarly disinclined to accept that it was a declining colonial power outpaced by larger-scale sovereign competition and managed to embark on some spectacular, and in hindsight misjudged, rocket-based defense projects of which Blue Streak, started in 1965 and scrapped in 1970 was probably the most notable example [11].

On a positive note, spending arguably disproportionate sums of money on missile development has produced a long-term legacy of rocket technology capability, both in France and the United Kingdom. France is a major contributor and economic partner in the European Ariane expendable launch system used to launch GSO and LEO satellites from Kourou in French Guiana with the rockets manufactured under the authority of the European Space Agency and Centre National D'Etudes Spatiales with Airbus as the prime contractor. The United Kingdom has a robust and relatively competitive near-space and deep space systems industry, BAE Systems being an example [12].

4.6 Rockets in the Rest of the World

In practice, a large part of present-day rocket launch technology is the product of intercontinental ballistic missile development work initially in Russia and the United States in the 1950s, China since the early 1960s [13], India since 1969 [14], Brazil since the 1970s [15], and more recently Israel, Iran, and North Korea since 2012.

More than 70 countries claim to have space programs including Malaysia [16], Indonesia [17], Egypt [18], Pakistan [19], and Egypt [20].

4.7 Indian Space Research Organization as an Example of New Emerging Nation-State Capabilities

On February 15, 2017, the Indian Space Research Organization [21] launched 104 satellites into orbit on a single rocket, setting a new world record. The rocket's main cargo was a 714-kg satellite for Earth observation but packaged with 103 nanosatellites weighing a combined 664 kg, with the smallest weighing 1.1 kg. Ninety of the nanosatellites are from a San Francisco company, Planet Inc., each weighing 4.5 kg, which will send Earth images back from space.

4.8 Brazilian Rockets and Their Sovereign Satellite Program

Brazil claims to be developing the technology to send domestically made satellites into space with locally made rockets by 2020. Visiona, a private sector joint venture with state-run telecom operator Telebras, and Embraer SA, the world's third largest commercial plane maker, are all part of an emerging rocket manufacturing and supply chain. The 2017 launch of the nation's first defense and communications satellite was a step towards this target of self-sufficiency with Thales SA and Ariane Space contracted to deliver the satellite into space supported by a large team of Brazilian engineers. The 5.8-ton geostationary satellite will provide broadband internet to Brazil and secure communications for military and government employees [22].

Brazil is the fifth largest country in the world with a total area of 8 million km^2, similar to Australia but including 55,455 km^2 of water. Indonesia is 2 million km^2. Africa is 30 million km^2. Continents trump countries in the bigness states. Although the present satellite build program has a limited amount of home-grown technology, the intention is to produce a microsatellite of 100 kg for launch into LEO at 1,000 km for missions such as deforestation monitoring, tracking hydraulic reservoirs, and monitoring 17,000 km of Brazilian border, although this is dependent on the state of the Brazilian economy.

4.9 China Long March Missiles

The latest big rocket from China is the Long March 5, the first of this big rocket series not to use solid fuel boosters. The liquid oxygen and kerosene-powered Long March 5 has a payload of 14,000 kg to GSO transfer orbit or 25,000 kg to LEO, close to the capability of the U.S. Delta 4 Heavy rockets. China placed a lander and rover on the Moon in 2013 and in 2016 sent two taikonauts, Chinese astronauts, into space to stay for 30 days on the Tiangong-2 space station. The next Long March 5 mission is scheduled to return to the Moon and the rocket, and the two planned successors (Long March 6 and Long March 7) are being designed to deliver an orbiter, lander, and rover to Mars by 2020.

4.10 European Rockets

Ariane 6 provides us with a well-documented example [23] of a contemporary European rocket being developed for use from 2020 onwards with participation from Austria, Belgium, Czech Republic, France, Germany, Ireland, Italy, the Netherlands, Norway, Romania, Spain, Sweden, and Switzerland. The vehicle is designed for launching payloads into geostationary orbit or intermediate orbit, MEO, or LEO/polar orbit up to a payload mass of 10.5 tons. The vehicle reuses engine components from Ariane 5 and is supposed to halve launch service costs.

Ariane 6 has a first stage using strap-on boosters based on solid propulsion (P), and second and third stages using cryogenic liquid oxygen and hydrogen propulsion (H) in a configuration known as PHH. The cryogenic main stage holds 150 tons of propellant, and the upper stage holds 30 tons. The design allows the rocket to be configured for two boosters (Ariane 62) or four boosters (Ariane 64) depending on the payload and orbit destination (it takes a lot more power to get to medium Earth or geostationary orbit). The four-rocket booster configuration will be used to double-launch commercial payloads of between 4.5 and 5 tons (5,000 kg).

Although Ariane remains a European-based development project, there is substantial cooperation between the European Space Agency and other countries, mainly due to the joint project work on the International Space Station [24]. Encouragingly, Russia and the United States agreed in 2017 to work on a "Gateway to Mars and the Cosmos" lunar orbit space station [25]. The project will almost certainly require further optimization of existing rocket technologies by the sovereign states involved.

The funding for Ariane would probably be more secure if Europe had a defense agency. The formation of a 5.5 billion Euro defense fund in July 2017 [26] is probably the first step towards this.

4.11 Solid Fuel versus Liquid Fuel

These sovereign rockets are either powered by liquid fuel, solid fuel, or a mix of both, although liquid fuel only is becoming dominant to make launches simpler and safer and more controllable and more environmentally friendly. Solid fuel rockets originally powered by gunpowder were invented by the Chinese in the thirteenth century. Modern solid fuel rocket engines use a wide range of chemical components. There are two general types of solid propellant, the first is known as a double-base propellant often based on nitrocellulose and nitroglycerine, the rocket motor is in effect an explosive device. Double-based propellant systems tend to be used in smaller rockets. The second type of propellant is known as a composite based on a mix of fuel and oxidized chemicals combined into a granular structure. The oxidizer is usually ammonium nitrate, potassium chlorate, or ammonium perchlorate, and the fuels are either hydrocarbons or derived from plastic [27]. These solid fuel combinations can be toxic, unstable, hazardous to handle, and particularly sensitive to mechanical shocks and changes in temperature. Once ignited, solid fuel rockets will burn to exhaustion and are commonly used as rocket boosters. Solid fuel boosters, for example, were used to launch the space shuttle into space.

Liquid fuel rockets use two separate propellants, a fuel and an oxidizer. The fuel can be kerosene, alcohol, hydrazine, or liquid hydrogen. The oxidizers include nitric acid, nitrogen tetroxide, liquid oxygen, or liquid fluorine. Liquid oxygen and hydrogen engines are often called cryogenic engines. Some of the best oxidizers such as oxygen and fluorine only exist as liquids at very low temperatures. Low temperatures can be hard to manage and maintain both before and during liftoff. Solid fuels and liquid fuels are both very capable of exploding when they should not. Liquid propellants can yield more energy than solid fuels, but require a much more complex engine and are therefore prone to mechanical and system failure.

4.12 The Rocket Men and Their Rockets

4.12.1 A New Generation of Space Entrepreneurs

Sovereign investment in rocket technology is not the only option. It can be observed that a new generation of aspiring space entrepreneurs are emerging with access to sufficient cash and borrowing facilities to develop private sector-based rocket and launch technologies. These technologies are technically and commercially competitive with legacy rocket systems with the potential to deliver a major reduction in delivery cost. SpaceX supports a mix of military and commercial payloads and is the first private-sector spaceship to have serviced the ISS having won the NASA International Space Station Cargo Resupply Services contract, which it shares with Orbital ATK, now a division of Northrop Grumman.

4.12.2 SpaceX Reusable Rockets and Other Innovations

SpaceX claims, almost certainly rightly so, to be the first company to develop a genuinely reusable rocket. Established in 2002 by Elon Musk, the founder of Tesla Motors, PayPal, and the Zip2 Corporation, SpaceX developed the Falcon 1 light-lift launch vehicle, the Falcon 9 medium-lift launch vehicle, Dragon, and Falcon Heavy, a heavy-lift vehicle capable of delivering over 50 metric tons into orbit.

A partial manifest excluding military sensitive missions includes Viasat, the U.S. Air Force, Telkom Indonesia, TELESAT, SSL, Sirius, SES, Northrop Grumman, NASA, KOREASAT, Inmarsat, Hispasat, Eutelsat, Arabsat, and Airbus [28]. Falcon 9 rockets are also being used by Iridium to launch their Next Constellation satellites; a total of 75 new satellites (66 satellites plus orbital spares) are being launched 10 at a time [29].

The rockets use liquid propulsion based on rocket-grade kerosene (RP-1), refined petroleum similar to jet fuel. This is combined with liquid oxygen (LOX) and generally described as a hydrocarbon engine. Nonexplosive pneumatic release and separation systems reduce orbital debris and are assembled on a horizontal production line to save cost and improve manufacturing safety and testability. The separation systems use kerosene from the propulsion system rather than hydraulic fluid in the hydraulic vector control systems in order to save weight and reduce system complexity.

These big rockets are immensely powerful. Falcon Heavy has 27 engines producing 22,819 kN (5,130,000 pounds-force [lbf]) of thrust at sea level and 24,681 kN (5,548,500 lbf) of thrust in vacuum. The second stage with 934 kN (210,000 lbf) of thrust delivers the rocket payload to orbit after the main engines have cut off and the first stage has separated to return to Earth. The engine can be restarted multiple times so that payloads can be placed on low Earth, GTO, or final geosynchronous orbit. The total available burn time is 397s.

The radio control systems onboard include S-band telemetry and video at 2,300 MHz to 2,300 MHz, a C-band radar transponder at 5,755–5,775 MHz, and seven separate radio systems using 11 radio channels. Iridium transceivers and tracking and GPS are used to facilitate the reusable stage landing and recovery process.

4.12.3 Price Lists and Payloads

Users can choose rocket launch options from a published price list [30]. For 2018, $62 million will secure a Falcon 9 rocket to take 22,800 kg to a LEO, 8,300 kg to a geosynchronous transfer orbit (GTO), and 4,020 kg to Mars.

Ninety million dollars will secure a Falcon Heavy launch, which will take 63,800 kg into LEO, 26,700 kg to GTO, and 16,800 kg to Mars (or 3,500 kg to Pluto).

Note that satellite operators can reduce launch cost by choosing a transfer orbit and then using the satellite propulsion system (hydrazine or an ion thruster) to lift the satellite to its final position. Ion thrusters have the advantage that they are powered from the solar panels on the satellite so the satellite effectively sails into space, but it can take 4 months to arrive at a final GSO orbital slot so this may not always be the most economic option.

The insertion orbits include LEO polar Sun-synchronous, also known as helio-synchronous, orbits. The Sun never sets on a helio-synchronous constellation. The satellites are used for Earth sensing or for Sun observation including early detection of unexpected sunspot events. A recurrence of a Carrington event (the last one was in 1859) would, for example, be significantly disruptive to both space and Earth-based computer hardware [31].

Customer payloads need to fit within a standard-sized container. The standard weight is up to 3,453 kg, and the premium is 10,886 kg. There is an expectation that the satellite or satellites will be delivered to more or less the required place in space typically no more than 10 or 15 km from the intended destination.

Note that electric satellites (satellites with ion thrusters referenced earlier and covered in more detail in Chapter 6) are capable of moving from an almost-there place in space to an exactly-there place in space. A transfer orbit will often be sufficiently near for an electric satellite to self-propel its way to its final destination, although, as stated, the time taken depends on the distance, the solar power budget, and the onboard ion propulsion system.

There is an additional requirement that the satellite ends up pointing in the right direction, preferably for communication satellites towards Earth or a particular part of Earth. This is known as the local vertical local horizontal (LVLH) orientation. SpaceX will also set the satellite spinning at a defined rate in order to realize the progressive pitch used to minimize interference with other satellite systems. A restart option available from the liquid propellant engine means that multiple satellites can be dropped off at different places in space.

4.13 Transporting Rockets to the Launch Site and Payload Launch Stresses

Environmental conditions for rockets and payloads are precisely specified for at least 3 weeks prior to a launch. Launch stresses on payloads at launch are also specified including heat, vibration, and noise and shock loads as individual stages separate. There is no point delivering a satellite into orbit if it falls apart on the way. Any RF systems on the satellite payloads also have to be capable of withstanding the RF flux density irradiated from the rocket during and after launch. Payloads must not transmit until released from the final stage.

4.13.1 Musk Mission to Mars 2024

Mr. Musk considers that a mission to Mars by 2022 or 2024 is achievable [32]. SpaceX has suggested that this mission might take precedence over their other entire rocket and launch system development work [33].

The ship is launched via the BFR booster, which then returns to Earth to be reused up to 1,000 times. During its journey to Mars, the ship docks with a refueling tanker, which also returns to Earth and then flies to Mars. On-site propellant production is then used to refuel the ship for its return to Earth. The fueling tankers and master ship are both designed to be reused up to 100 times.

Lockheed has a similar mission proposal [34]. In Figure 4.4 it is on the Red Planet, Mars.

4.13.2 Mr. Bezos and Blue Origin

Jeff Bezos, the founder of Amazon, also has his own rocket company and reusable launch solution called Blue Origin [35] (Figure 4.5). He intends to invest $1 billion per year in the venture from sales of Amazon stock [36].

Figure 4.4 Lockheed Martin mission to Mars. (Thanks to the Lockheed Martin Company.)

Figure 4.5 Blue Origin reusable rocket. (Courtesy of the Blue Origin Image Library.)

Blue Origin has two rocket engines under development and test, a liquid hydrogen fueled-engine developing 110,000 lbs of thrust (over a million horsepower) as a first-stage rocket and a liquid oxygen and liquefied natural gas engine producing 550,000 lbs of thrust for deep space missions. The liquefied gas is also used to pressurize the rocket's propellant tanks. Unlike kerosene, liquid gas produces no soot products making engine reuse easier. The engine has the ability to restart multiple times. The launch engine can be throttled back to a thrust of 20,000 lbs to provide sufficient control for landing within a few meters of a designated launch area.

As with all rocket engines, the material challenge is to manage the extreme temperature gradients within the rocket engine with the liquid hydrogen needed to be maintained at −423°F in close proximity to combustion temperatures of 6,000°F and to get systems such as the turbo pumps to a high level of reliability and controllability.

Blue Origin is part of the United Launch Alliance [37], and the engines have been chosen by the alliance to power the next generation of Vulcan launch vehicles. The United Launch Alliance is a 50:50 joint venture between the Boeing Company and Lockheed Martin formed in 2006 to provide lower cost delivery services for U.S. government missions. The alliance combined the two rocket development teams, Delta and Atlas, which between them had supported 50 years of U.S. space effort, a total of 1,300 missions. The alliance today employs just over 3,000 engineers on subspace, near space, and deep space rocket-related development for work for the U.S. Department of Defense, NASA, the National Reconnaissance Office, and the U.S. Air Force.

4.13.3 My Rocket Is Bigger Than Your Rocket

The biggest rocket that Mr. Bezos is building is the New Glenn (named after John Glenn, the pioneering U.S. astronaut). The rocket is 82m tall with a capacity to lift 45 tons to LEO or 13 tons to geostationary transfer orbit with a maiden flight planned for 2019. The rocket is designed to be able to be reused 100 times, coming back to Earth and landing on a sea barge.

New Glenn is one of the biggest but not necessarily the biggest of this new generation of heavy lift booster rockets (Figure 4.6). The SpaceX Falcon Heavy will have the capacity to deliver 53 tons to low earth orbit and NASA is working on a new space launch system with a 70-ton capacity.

4.13.4 Mr. Branson and Virgin Galactic

Richard Branson is promoting Virgin Galactic [38] for space tourism, the world's first commercial space airline, but also for space debris cleanup and low-cost LEO small-satellite delivery using the Launcher One system [39] (Figure 4.7). Founded in 2005 by Richard Branson and Bert Rutan, the founder of Scaled Composites [40], the project has suffered a couple of high-profile failures, but seems to be making progress towards a regular flight manifest.

The Virgin Group is also an investor in OneWeb and Boom Technology. Boom Technology [41] is developing a new supersonic passenger transporter as a successor to the Concorde.

Sending celebrities into space is high-risk in terms of public relations, but if space travel can be delivered at a similar safety level to conventional commercial airlines, then insurance costs would be significantly reduced.

Note also the close coupling between aircraft design and development, spacecraft design and development, and potentially air travel and space travel. For Virgin, this includes an initial lift of the space vehicle on a very large but conventional airplane (The Strato Launcher).

This initial lift (an air lift to 50,000 ft and then an air launch) combined with the relatively low apogee of the flight plan (110 km, 10 km beyond the Karman limit) means that a smaller thrust engine can be used on the space craft. The Virgin vehicle engine uses a liquid nitrous oxidizer with thermoplastic polyamide, essentially nylon, as the solid fuel component of the propellant to produce 60,000 lbf of thrust (270 kN).

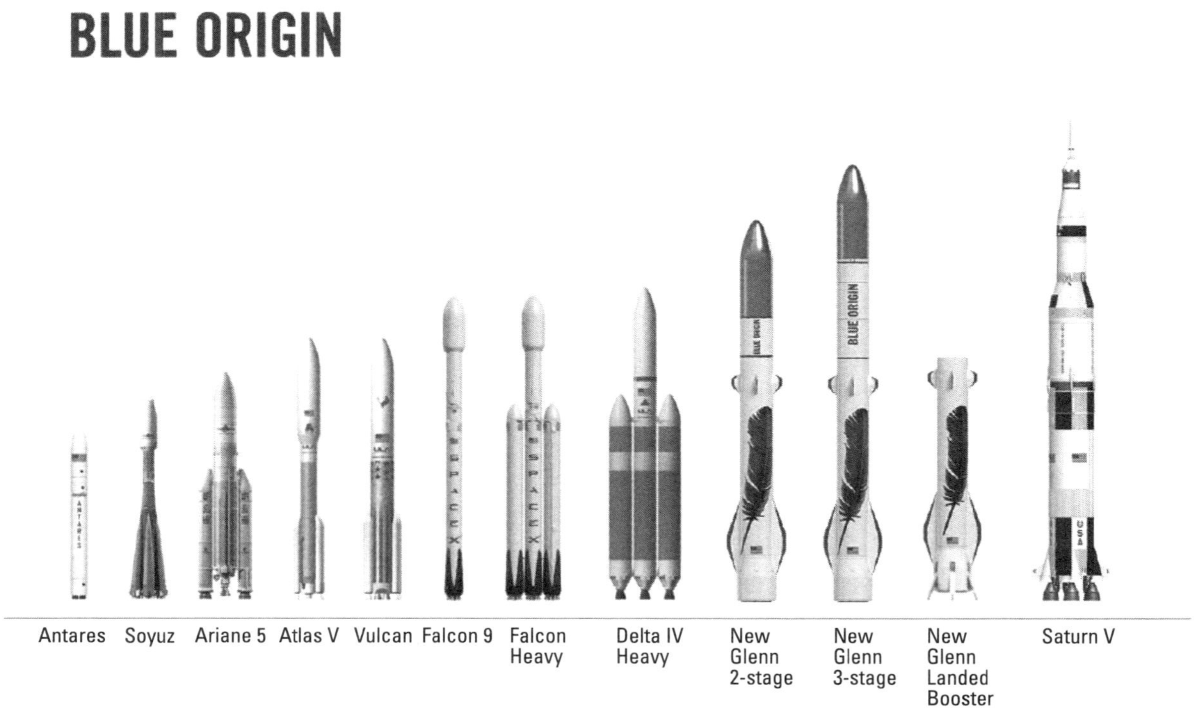

Figure 4.6 Blue Origin Rocket New Glenn's size comparison with other rockets.

Figure 4.7 Virgin Galactic, the world's first commercial space airline.

4.13.5 Small Rockets: The Ki-Wi Way

This brings us to the topic of small rockets and Rocket Lab USA [42]. Rocket Lab was set up by Peter Beck in 2006 and includes Lockheed Martin as an investor and K1W1, a New Zealand venture capital fund [43]. The rocket makes extensive use of carbon composite materials including a carbon composite mix for the fuel tanks that is compatible with liquid oxygen and a carbon composite fairing. The oxygen/kerosene pump engine uses battery-driven electric propellant pumps and is three-dimensionally (3-D) printed with a print time of 24 hours, a process now commonly described as additive manufacturing.

Nine of these engines in the first stage produce a liftoff thrust of 162 kN (34,500 lbf) with a peak thrust of 192 kN (41,500 lbf) and a burn time of 303s. The second stage uses a modified version of the same engine optimized for performance in vacuum conditions with a burn time of 333s. The 17-m mini rocket is designed to launch a nominal payload of 150 kg into a 500-km Sun-synchronous orbit. The company has its own launch site in Mahia, New Zealand.

Rocket Lab is positioned to service the CubeSat market for low-cost space research projects with a suggested price of $50,000 to $90,000 for a 4 inch by 4 inch CubeSat or $180,000 to $200,000 for a 12 inch by 12 inch CubeSat.

4.13.6 Micro Spacecraft Launchers

Rocket Lab is one of a number of start-ups with ambitions to develop low-cost space delivery systems. One example is Vector Space Systems [44], which has a rocket similar to Rocket Lab (carbon fiber and similar propulsion system) and a similar market target of small payloads, either 160 kg or 66 kg but launched from a mobile launcher. The first stage of the rocket is designed to be reusable. They claimed they just need an empty concrete car park as a launch site. The company has also worked on nozzle control systems to maximize thrust at all altitudes during a launch. These are individually small technical and mechanical improvements but taken together deliver potentially significant improvements in launch efficiency and are examples of a newly emerging space supply chain that will undertake rocket development to order, the Garvey Spacecraft Corporation is another example [45].

4.13.7 How Far Away Is Space?

The start of space is generally accepted as the Karman line named after the Hungarian scientist Hr. Theodore Von Karman (1881–1963). It is the boundary at which aerodynamic forces are no longer consequential and at which orbital centrifugal force rather aerodynamic lift becomes the dominant mechanism for maintaining a defined altitude. The boundary limit was finally agreed in the 1950s.

In between the Earth and space is the troposphere ranging in thickness from 8 km at the poles to 16 km above the equator, bounded by the tropopause, a boundary marked by stable temperatures. The troposphere is where most of the world's weather takes place including hurricanes and thunderstorms, both of which can be disruptive to rocket launches and launch facilities.

Above the troposphere is the stratosphere between 12 and 50 km. The stratosphere defines a layer in which temperatures rises with altitude. This rise in temperature is caused by the absorption of ultraviolet (UV) radiation from the Sun by the ozone layer. This creates stable atmospheric conditions, with minimal air turbulence and strong steady horizontal winds (the jet stream) and is therefore ideal for aircraft. The thin air at the top of the stratosphere is close to 0°C.

The Mesosphere is from 50 to 80 km above the Earth's surface. It is separated from the stratosphere by the stratopause and from the thermosphere by the mesopause. Temperatures in the mesosphere decrease with altitude to about −100°C.

The thermosphere is the outer layer of the atmosphere, separated from the mesosphere by the mesopause. Within the thermosphere, temperatures rise continually to well beyond 1,000°C due to the heat energy absorption from the Sun. However, the air is so thin that it feels very cold. The ionosphere extends up to 600 km and is subject to substantial ionization from UV radiation absorption and from high-energy particles from the Sun and cosmic rays. The radiation levels in the troposphere can be damaging to electronics and some components and systems have to be radiation-hardened.

The Van Allen belts [46] have particularly high radiation and significant space weather events that have to be factored into mission risk mitigation plans. Earth's two main belts extend from an altitude of 500 to 58,000 km. Their discovery is credited to James Van Allen [47] (1914–2006).

4.13.8 Near Space versus Deep Space

As a rule, anything described as near space is within the Earth/Moon orbital system. Anything further away than the Moon is defined as deep space

4.13.9 How Long Does It Take to Get There?

To get into near-Earth orbit, a first-stage powered flight lasts approximately 3 minutes, and second stages burn for an additional 5 to 6 minutes to reach initial orbit. If you ran there, it would take about 12 hours (the average time that it takes an ultra-runner to complete a 100-km ultra) (Figure 4.8).

Figure 4.8 A 12-hour run into space. (I am on the left at age 100 after running 100 km.)

It takes 3 days to get to the Moon and about 7 months to get to Mars via the Hohmann transfer orbit [48]. The Hohmann transfer orbit is named after the German space scientist Walter Hohmann (1880–1945) [49].

The spacecraft Voyager 1 and Voyager 2 [50] (both launched in 1977) have taken 40 years to reach the edge of the solar system, a 12-billion-mile journey so far, and are making their way through the heliopause towards the Oort clouds. They will reach the Oort clouds in 300 years and exit the other side in 30,000 years [51]. Space is a big place and getting bigger all the time.

4.14 The Impact of Big Rocket Innovation on High-Count LEO Power Budgets, Capacity, Throughput, and Space Constellation Economics

Figure 4.9 shows an estimate from Nokia Networks of expected data density in a 5G network calculated for sub-6-GHz networks, centimeter-band networks, for example, 12-GHz Ku-band and 28-GHz (or 26-GHz) Ka-band networks, and millimeter-band networks (V-band at 40 and 50 GHz and E-band either side of the 77-GHz automotive radar band).

We revisit this topic again in Chapter 10, but for the purposes of our present topic (launch technology), it is useful to consider how much RF capacity and RF power can be delivered into space at what cost. The cost (satellite cost + launch cost + though-life operational cost) together with other metrics such as latency will determine the amount of terrestrial bandwidth that can be serviced from space.

The typical launch payload in the past has been of the order of 10,000 kg. The next generation rockets being built for Mars Missions can lift more than 60,000 kg into LEO. This would mean that one rocket could lift 120 500-kg satellites into LEO in one go, which would mean 1,400 satellites in a year on a monthly launch cycle or four times that on a weekly launch cycle.

This suggests that SpaceX could provision a 4,000-satellite constellation in space within a 1-year launch window. If each satellite had 500W of RF power

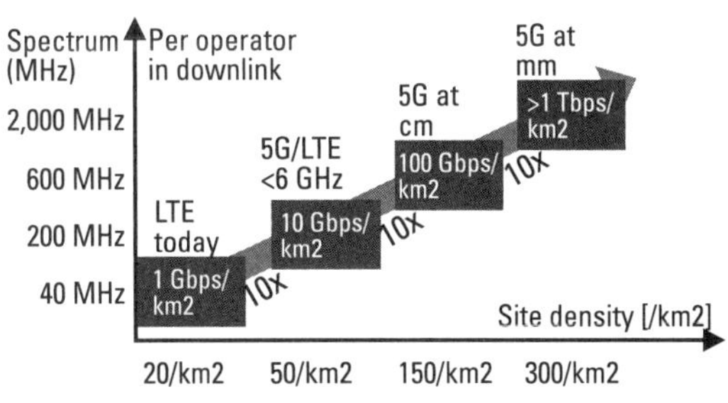

Figure 4.9 Data density and site density in a 5G network.

and 1 Tbps of throughput, then this would produce a network with 2 MW of available RF downlink power with a throughput of 4 petabits. This power and throughput could be scaled across small cells (2-km radius) to large cells (horizon to horizon) with multiple beams servicing individual users or IoT devices within those cells. The satellites have a 20-year life expectation, access to free energy, and no site costs. This looks attractive, but there are other unexpected additional costs.

4.15 The Impact of Launch Reliability on Insurance Cost

One of these additional costs is insurance and the cost of insurance is as least partly determined by launch reliability. The space sector insurance market was established in the 1960s but little used as the satellites and the satellites flown on them were owned by the governments of the countries that launched them and operated by government agencies such as NASA and the European Space Agency (ESA).

The market began to grow in the 1980s as satellites became more sophisticated and the consequences of a launch loss or post partition loss (the bit between being separated from the rocket and arriving at the final orbital destination) or in orbit loss became financially and politically more painful. The Space Shuttle *Challenger* disaster in 1986 provided a tragically high profile reminder of the risks inherent in the launch industry [52].

The general growth of the insurance market meant that funds were available for underwriting risks where launch reliability statistics were available for launch failures and interim and final orbit failure. The ability for a rocket and its satellite to do damage on the way to space or inadvertently on an unexpectedly early return meant that traditionally self-insured programmes began to purchase premiums, which could require as much as $400 million of coverage. The premiums would vary depending on the risk/reward ratios of other parts of the insurance economy.

Just over 50 years ago, the Torrey Canyon [53] had substantially increased costs to the industry. These were reflected in increased premiums not only for the maritime sector but for all insurance. More recently, Hurricane Katrina had a similar industry-wide impact with over $40 billion paid out to cover damage costs [54].

In 1998, solar storms tipped the satellite insurance industry from a 20-year run of profits to a loss of over $1 billion, and the sector became a pariah until other disasters became more visible. Before the solar storm, rates for the satellite sector were around 15% to 16% of insured value and typically offered coverage for launch and a year of on orbit coverage. After the storm, rates actually reduced on the basis that lightning does not strike twice, or at least for solar cycles, for another 11 years, so by 1999, rates of well under 10% could be negotiated, which included 5 years of in-orbit protection.

The insurance industry is full of these counterintuitive pricing effects. A disaster prompts actuarial teams to take a closer look at risk/reward ratios in other sectors and funds get reallocated. In a market dominated by two or three large insurers out of a total pool of 15 or 20, a follow-them effect ensures a relatively dramatic reallocation of funds and the additional competition and need for market share drives premiums down.

This is offset by a "this cannot happen again for a while" mentality, although in the case of cyclical risk, as in the solar storm cycle, this can be quantified and calculated.

The rough rule of thumb used to be that there was about a 1 in 10 risk of losing a satellite during the launch process and a 1 in 20 risk of a satellite failing in the first 6 to 12 months of operation. The in-orbit hazards included damage from electrostatic discharge caused by solar activity or from plasma clouds formed from meteoroid showers, direct impacts from meteoroids, probably the cause of the failure of the ESA satellite Olympus in 1993. Solar storms can cause damage either electrically or for LEO satellites, by causing warm air to rise. This introduces additional drag and means the satellites have to use fuel to maintain their orbital altitude. GSO satellites are more vulnerable to meteor damage. Meteors begin to burn up at LEO altitudes. Last but not least, there is the risk of damage from space debris.

Some of these risks have decreased over the past 30 years or can be managed more effectively. For example, electric satellites (covered in Chapter 5) have more or less solved the problem of station-keeping-related fuel depletion.

Some of the risks are increasing or can be assumed to increase in the future. The general assumption, for example, is that damage from space debris will increase over time.

This can either cause catastrophic failure or just degrade performance. In February 2009, a defunct Russian military satellite collided with an Iridium 33 satellite destroying both objects and creating a bit of a mess and there have been reports of subsequent collisions [55].

More common are the wear-and-tear issues, for example, the fogging of solar panels due to debris abrasion (covered in Chapter 5). The grab handles of the ISS have had to be replaced due to becoming razor sharp from continuous microscopic impacts.

The question is whether rockets are becoming more reliable and if so by how much and how quickly. On the positive side, materials and manufacturing innovation including more accurate computer based preflight and post flight testing combined with larger production volumes and more automated production processes (humans are generally less reliable than machines when it comes to making more machines) should all combine together to reduce launch failure. The recovery of first stage rockets (SpaceX and Blue Origin) will also be yielding data on materials performance, which should result in additional design optimization.

This issue of reliability can be a barrier to entry for new rocket manufacturers competing against legacy launch systems such as Ariane (European), Delta (U.S.), Long March (Chinese), and Proton (Russian) launch systems. The Delta rocket failed on its first launch and then had 23 successful launches before another failure. Because of the initial failure, the rocket had to be launched 5 times before it achieved an 80% success rate. The development of the first generation of rockets by SpaceX between 2006 and 2009 started with three launch failures, which produced a reliability rating of 0% but that is now ancient history and SpaceX, partly due to its NASA contract to supply the ISS (2008 onwards), now successfully competes more than adequately with the traditional contracting community including the mighty Boeing and Lockheed Martin Corporation.

The cost benefit of reliability is also partly determined by whether space insurance rates are high or low. When rates are low, there is about a 1:1 relationship between reliability and insurance premium. When rates are high, this increases to a 1:4 relationship, the safe bet becomes the preferred option.

There is an ongoing political debate as to whether governments provide indemnity caps to launch companies. The historic justification for this has been the military and strategic importance of having a sovereign nation launch capability. The U.S. Space Renaissance Act, with the express purpose of establishing the United States as the preeminent "space faring nation" [56], provides an example of the continuing political importance of the space sector.

Last but not least, if the life time of satellites continues to increase (Chapter 5 shows why this is happening), then insurance premiums will become a lower overall cost component in the lifetime cost of owning and operating a satellite constellation.

4.16 Summary

Unsurprisingly, 60 years of satellites have been closely coupled to 60 years of rocket development with the first 40 years dominated by large rockets designed with the purpose of delivering nuclear warheads around the world or astronauts and military systems into space. These were essentially sovereign nation rocket development programs or for Ariane, a multicountry (EU) collaborative project.

These legacy expendable large rockets are still used today for satellite launches but are gradually being replaced by reusable large rockets developed by the private sector, with SpaceX and Blue Origin New Glenn being significant new players in the satellite delivery business. There are also small rockets developed by private-sector companies and small reusable space vehicles, Virgin Galactic being the one notable example. Table 4.1 summarizes the relative size of these big and small rockets.

There is still a political appetite for developing and building large sovereign nation rockets. The new space launch system being developed by NASA is based on a rocket that is over 120m tall, bigger even than Saturn 5, the rocket developed and built to take astronauts to the moon. Since its retirement in 2011, the United States has had no ability to launch its own astronauts into space and has been buying space on Russian Soyuz spacecraft, a rocket designed in the 1960s. These massive rockets can either take a smaller payload a long way, for example, to Mars, or use all that power to deliver very large loads into LEO, MEO, and GEO.

However, the initial flight of the new NASA launch vehicle is budgeted to cost at least $7 billion and it is open to debate whether the private space sector could provide the same or similar capability at a lower cost. NASA counterargues that it can use private-sector facilities to reduce cost. For example, the engine testing is being done at a facility owned by Orbital ATK, now a subsidiary of Northrop Grumman.

For the 5G and communications satellite community, the main conclusion that can be drawn from this chapter is that the rocket business has dramatically changed in the last 10 years both in terms of technology positioning and commercial positioning.

The big technology change is the shift to reusable rockets, but there are many smaller incremental technology improvements that are making space delivery more reliable and therefore less expensive including lower insurance costs. Note that it is not just the rocket that needs to be insured but also the cargo in

Table 4.1
Big Rockets and Small Rockets

Big Reusable Rockets		**Small Rockets**	**Small Reusable Space Vehicles**
SpaceX Falcon Heavy	Blue Origin New Glenn	Rocket Lab	
70m tall	82m tall	17m tall	Big enough for 10 people
Liquid fuel	Liquid fuel	Liquid fuel	Liquid nitrous oxidizer with thermoplastic polyamide solid fuel propellant
>200,000 kN	>200,000 kN	200 kN	200 kN
50 tons to LEO orbit	50 tons to LEO orbit	250 kg to 500 km Sun-synchronous	10 people to 110 km

the hold, a satellite, or multiple satellites. The satellites can be more expensive to insure than the rocket. Reducing launch failures therefore has a significant impact on overall cost.

The use of carbon composite materials, 3-D printing, new welding techniques (covered in Chapter 8), and optimized horizontal production lines and improved test procedures are all having a positive impact on space delivery economics. Commercially, the demand for military payloads remains robust and helps to provide cross-amortization opportunities for rockets that can take mixed military and commercial and consumer payloads into space that range in size and weight from a grapefruit to a large double-decker bus. These technology and commercial innovations are having an equally beneficial impact on satellite cost and performance.

References

[1] http://www.ssgreatbritain.org/your-visit/collection-stories/brunel-aboard-great-easternhttp://www.ikbrunel.org.uk/ss-great-eastern.

[2] http://www.astronautix.com/n/newtonsorbitalcannon.html.

[3] https://www.wired.com/2011/05/0525arthur-c-clarke-proposes-geostationary-satellites/.

[4] https://www.space.com/20063-hermann-oberth.html.

[5] http://www.v2rocket.com/start/makeup/motor.html.

[6] https://www.space.com/19944-robert-goddard.html.

[7] http://www.russianspaceweb.com/rockets.html.

[8] https://www.nasa.gov/centers/marshall/home/index.html.

[9] https://www.nasa.gov/centers/marshall/news/news/releases/2017/nasa-tests-first-3-d-printed-rocket-engine-part-made-with-two-different-alloys.html.

[10] http://www.world-nuclear.org/information-library/country-profiles/countries-a-f/france.aspx.

[11] http://www.nationalarchives.gov.uk/films/1951to1964/filmpage_rocket.htm.

[12] http://www.baesystems.com/en-uk/capability/space-systems.

[13] https://www.space.com/22743-china-national-space-administration.html.

[14] http://www.isro.gov.in/.

[15] http://www.aiab.org.br/site-ingles/indutria-aeroespacial.asp.

[16] http://www.angkasa.gov.my/.

[17] http://www.iafastro.org/societes/indonesian-national-institute-of-aeronautics-and-space-lapan/.

[18] http://www.narss.sci.eg/.

[19] http://suparco.gov.pk/webroot/index.asp.

[20] http://www.narss.sci.eg/divisions/view/5/%20Space%20Sciences%20and%20 Strategic%20Studies/16/Egypt%20Space%20Program.

[21] http://isro.gov.in/update/15-feb-2017/pslv-c37-successfully-launches-104-satellites-single-flight.

[22] http://www.reuters.com/article/us-brazil-satellite/brazil-ramps-up-domestic-space-satellite-rocket-programs-idUSKBN16T2BD.

[23] http://www.esa.int/Our_Activities/Space_Transportation/Launch_vehicles/Ariane_6.

[24] http://www.esa.int/Our_Activities/Human_Spaceflight/International_Space_Station/Where_is_the_International_Space_Station.

[25] http://www.dailygalaxy.com/my_weblog/2017/09/usa-russia-lunar-space-station-announced-a-gateway-to-the-cosmos.html.

[26] http://europa.eu/rapid/press-release_IP-17-1508_en.htm.

[27] https://history.nasa.gov/conghand/propelnt.htm.

[28] www.spacex.com/missions.

[29] https://www.nasaspaceflight.com/2017/06/spacex-falcon-9-iridium-next-2-launch/.

[30] http://www.spacex.com/about/capabilities.

[31] http://www.history.com/news/a-perfect-solar-superstorm-the-1859-carrington-event.

[32] https://www.newscientist.com/article/2149003-elon-musks-new-plans-for-a-moon-base-and-a-mars-mission-by-2022/.

[33] https://www.space.com/38313-elon-musk-spacex-fly-people-to-mars-2024.html.

[34] http://www.lockheedmartin.com/us/ssc/mars-orion.html.

[35] https://www.blueorigin.com/.

[36] https://www.theverge.com/2017/4/5/15200102/jeff-bezos-amazon-stock-blue-origin-space-travel-funding.

[37] http://www.ulalaunch.com/.

[38] http://www.virgingalactic.com/.

[39] http://aviationweek.com/space/new-virgin-orbit-formed-lead-smallsat-launch-vehicle.

[40] http://www.scaled.com/.

[41] https://boomsupersonic.com/.

[42] https://www.rocketlabusa.com/electron/.

[43] https://www.crunchbase.com/organization/k1w1.

[44] https://vectorspacesystems.com/

[45] http://www.garvspace.com/.

[46] https://www.nasa.gov/mission_pages/sunearth/news/gallery/ScienceCover.html

[47] https://www.nasa.gov/vision/universe/features/james_van_allen.html.

[48] https://www.mars-one.com/faq/mission-to-mars/how-long-does-it-take-to-travel-to-mars

[49] http://pioneersofflight.si.edu/content/walter-hohmann-0.

[50] https://voyager.jpl.nasa.gov/.

[51] https://www.nasa.gov/mission_pages/voyager/voyager20130912f.html.

[52] https://www.space.com/18084-space-shuttle-challenger.html.

[53] https://www.theguardian.com/environment/2017/mar/18/torrey-canyon-disaster-uk-worst-ever-oil-spill-50tha-anniversary.

[54] https://www.insurancehunter.ca/blog/impact-hurricane-katrina-insurance-industry.

[55] https://spaceflightnow.com/2015/01/22/did-two-more-iridium-satellites-collide-with-space-debris/.

[56] http://spacerenaissanceact.com/.

5

Satellite Technology Innovation

5.1 The Power of Power

This chapter follows a similar trajectory to Chapter 4, and the common theme is power. We discussed the positive impact of the planned Mars missions on propulsion system development. For satellites, we will find that the planned Mars missions are having a similar positive impact on power system development.

We finished Chapter 4 with a comparison of small rockets and large rockets and the trade-off between going a long way, to Mars and beyond, with a relatively small payload, or a short way, a run into LEO, with a large payload. We highlighted that the development spending on vehicles capable of fulfilling deep space missions beyond the Moon is having a directly beneficial effect on the lifting power available for new generation communications satellites. These can either be huge satellites weighing thousands of kilograms or tiny satellites weighing a few kilograms in weight. Medium-sized satellites such as Iridium (around 500 kg) can be launched 10 at a time from these big rocket systems; alternatively, 100 small satellites can be packed into a single nose cone and spread into LEO. Orbcomm, the VHF constellation referenced throughout the book, has 31 satellites in LEO at 775 km. Seven replacement satellites were launched in 2017 by SpaceX as a secondary payload.

The common denominator is the amount of thrust available and the economic cost of providing that thrust either once (with expendable rockets) or multiple times (reusable rockets). The ambition is to reuse rockets up to 100 times. This will substantially change the economics of delivering consumer, commercial, and military payloads into space.

We also referenced the close coupling between military spending on rockets including intercontinental ballistic missile development and present-day commercial rocket systems particularly in terms of liquid fuel engine development and guidance systems and added in the impact of materials innovation, in particular the use of carbon composite materials and new manufacturing techniques. There are many crossover opportunities with the global aircraft manufacturing industry with aircraft such as the Boeing Dreamliner making extensive use of carbon composites, for example, to reduce weight and improve performance.

However, in the specific context of satellites, we are not talking just about propulsive power but a combination of propulsive power and processing power. The birth of the satellite industry 60 years ago was only made possible by advances in component technology including the transistor in the early 1950s. The development of satellite capability since then has been driven by the development of the microcontroller in the 1970s, the digital signal processor in the 1980s, and low-cost, high-performance memory in the 1990s. As we shall discuss in Chapter 9, the combination of these innovations is opening up a host of new space-based opportunities including servers in the sky, above the cloud computing and new dot.space business models.

Crucially, these new satellite constellations will coshare spectrum with existing MEO and GSO satellites. They achieve this through angular power separation which requires the satellites to roll as they fly towards the equator (progressive pitch control).

Solar panels have a number of disadvantages. They are vulnerable to damage, create additional space debris in the event of a collision, and make it harder to dock multiple satellites together. Sometimes known as buddy SATs, docking multiple satellites together at the same orbital position is a useful potential option for scaling bandwidth particularly for GSO constellations constrained by the need to have 2° of separation between orbital slots.

Solar panels are relatively inexpensive compared for example to nuclear power sources and more environmentally benign, although if additional launch weight and volume cost is taken into account, the cost difference reduces.

If radioactive power sources could be reduced in cost, there would be a new market for space-optimized, weight-optimized, small, nuclear power sources with a unit volume potential of at least 10,000 satellites, a market volume many orders of magnitude larger than any existing maritime or terrestrial nuclear power source application. Nuclear power in space could also help meet the energy efficiency and carbon footprint targets for 5G terrestrial networks, a topic to which we will return in later chapters.

Last but not least, alternative power sources could provide the additional power needed to keep low Earth satellites in orbit for longer and would overcome the problem of solar panel performance degradation.

5.2 The Sun as a Source of Power

Close to Earth, in LEO, MEO, and GSO, there is sufficient sunlight to power most communication satellites. Solar energy is measured in W/m^2, energy per unit area. Outside the Earth's atmosphere, this is roughly 1,350 W/m^2. On the Earth's surface, this is a maximum of around 1,000 W/m^2 for extreme desert areas. In addition, on Earth, there are night and day cycles, so there is more solar energy in space.

5.2.1 Solar Panel Efficiency

Consumer-grade solar panels that you may have at home typically have an efficiency of 15%. Solar panels for use in space use optimized manufacturing techniques based on germanium rather than silicon with cells that have multiple p-n junctions that capture different portions of the energy spectrum. These panels can achieve efficiencies approaching 30% [1]. The systems also depend on the ability to manufacture space-qualified cover glass to protect large arrays of solar panels from impact damage. The Opportunity Mars Land Rover has just finished 11 years of exploring the surface of the planet from the original landing sight. The original expectation was that the solar panel arrays would not survive for longer than 90 days. The solar panel supplier for many of these deep space projects is a subsidiary of the Boeing Corporation [2].

5.2.2 The International Space Station as an Example of Big Solar Panels in LEO

The poster child of large solar arrays is the ISS orbiting in a LEO at 350 km with four sets of solar arrays, each 33m long and over 12m wide producing 200 kW of electricity. Together, the arrays contain a total of 275,0000 solar cells and cover an area of about 2,500 m^2, more than half the area of a football field.

5.2.3 Satellite Power Requirements

A large GSO satellite will typically have an onboard power requirement of 15 kW, a medium-size LEO such as an Iridium NEXT constellation satellite will have an onboard power budget of about 500W, and a CubeSat has to be happy with a few milliwatts (see Table 5.1). The solar arrays on these platforms have to survive many years in space and have to be designed so that any efficiency losses through life are minimized.

The Inmarsat Ka-band GSO satellites, for example, have solar panel arrays with a wing span of 33.8m with ultra-triple-junction gallium arsenide solar cells that generate 15 kW of power at start of service and 13.8 kW by the end of life (15 years). The panels also power the xenon ion propulsion system. This is a massive beast with a main body the size of a double-decker bus and a launch

Table 5.1
Typical Satellite Power Requirements

Picosatellites (CubeSats)	Nanosatellites	Microsatellites	Macrosatellites
<1 kg	<10 kg	<500 kg	≥500 kg
Milliwatts	Tens of milliwatts	Hundreds of watts	Kilowatts

weight of over 6,000 kg. Eighty-nine Ka-band user beams are generated by two transmit and receive aperture antennas with an additional six steerable on-demand spot beams (Figure 5.1).

5.2.4 The Power of Solar Power and What It Is Used For

Solar power is used in these near-Earth satellites to keep the onboard processors happy, in terms of both their energy requirements and their ambient temperature, to power the systems that determine the orbital altitude and pitch and yaw control of the satellite, typically with momentum motors and to provide RF power for the onboard transceivers that are needed both for telemetry and to provide communications to and from Earth. As a rough rule of thumb, the split between the RF power requirement and onboard baseband processing is 50:50.

Figure 5.1 Inmarsat GSO satellite. (Image courtesy of Inmarsat.)

5.3 The Importance of Satellite Power Efficiency

This explains why RF power efficiency is critically important for satellite transmitters and why power efficient modulation schemes are used together with closely controlled multiplexing with minimum amplitude modulation (AM) variation. There will be a trade-off depending on whether the satellite is functioning as a repeater or a relay. A relay, generally described as a bent pipe satellite, takes an uplink signal, amplifies it, and sends it back to Earth. This minimizes onboard processor delay and power consumption, but any noise on the signal is also amplified. A relay demodulates the uplink (a multiplex of many users separated in the time domain) and decodes then codes and modulates the downlink. This will use more power, but the residual bit error rate will be lower, so if well implemented it should realize a better efficiency on a user bit-throughput basis (or goodput as our internet friends insist on calling it). There will also be some difference in power budget distribution depending on whether intersatellite switching is used. Iridium intersatellite switch in Ka-band, SpaceX, and LeoSat are proposing to intersatellite switch with optical transceivers. On balance, it is more complicated to have switching onboard the satellite, which means that more processing power will be consumed and there are more functions that can go wrong due to, for example, radiation damage of microprocessors or RF components, but the latency gains can be significant (see Chapter 4). Intersatellite switching also reduces the number of Earth stations needed to support the constellation. This can have a substantial impact on terrestrial asset costs. In general, therefore, it pays to put more processing in space particularly if the processing is power efficient.

5.4 Electric Satellites Using Ion Propulsion Systems

Solar panels can also be used to provide the power to fly satellites from interim orbits to their final orbit destination. This is often described as orbit raising and the satellites are described as electric satellites. Their purpose is to reduce launch costs but also to optimize through life station keeping, which prolongs the life of the satellite. This is because satellites relying on hydrazine thrusters must be deorbited before the hydrazine runs out.

Electric satellites (like the one shown in Figure 5.2) use ion thrusters as a propulsion system. Chemical rockets, as we documented in Chapter 4, can produce 200,000 kN of thrust but use a large amount of fuel to generate that thrust. The thrust efficiency is a function of exhaust velocity. Liquid oxygen and hydrogen produce an exhaust velocity of about 5,000m per second.

Ion thrusters take a noble gas, which is inert and chemically unreactive, such as xenon and strip or add electrons to produce plasma, which is then accelerated with an electric or magnetic field producing an exit velocity out of the

Figure 5.2 Airbus electric satellite. (Credits Airbus.)

back of a thruster about 10 times that of a chemical propellant. Xenon is easily ionized and has high atomic mass and high storage density. However, unlike chemical propellants, the gas does not have power of its own to release and the thrust is therefore a function of the amount of electric power available.

Ion propulsion technology was developed in the late 1950s and first tested in space in the early 1960s and today is routinely used in deep space missions and to keep geosynchronous satellites at their correct location. Considerable work is being done to increase the output and efficiency of ion thrusters [3]. Power outputs range from a few watts to kilowatts [4]; 1 kW equals 1,000N per second [5].

Boeing introduced what it claimed was the world's first all-electric satellite in 2015. The satellite uses three Hall Effect [6] plasma thrusters to get from a transition orbit to a final orbit position or to change orbit while in service. For example, in a Leo constellation, a satellite could be kept in a reserve orbit and then flown up to operational orbit when needed. The Boeing ion thruster is rated at 5 kW [7].

Europe has its own electric satellite project (Electra [8]) supported by the satellite fleet operator SES, Swedish satellite manufacturer OHB-SE [9] and the European Space Agency aimed at satellites with a weight of 3,000 kg or less at launch (similar to the Boeing 702sp satellite).

The Hispasat small GEO satellite is another example of an electric satellite [10]. Hispasat 36W-1 uses chemical propulsion to climb into final geostationary position after separation from the rocket and then an electric propulsion system to manage station keeping for its anticipated 15-year lifespan.

Ion thrusters are therefore an additional example of deep space technology originally developed over 50 years ago and refined over 5 decades, being applied to near-Earth satellites to reduce launch payload costs and though life costs (by extending the lifetime of the satellite).

5.5 What Happens When the Sun Stops Shining?

Further out into space, the Sun becomes progressively less useful as a power source (the Oort clouds are a particularly gloomy place in space) but any journey away from the Sun or in the shadow of other planets can be compromised by a shortage of solar energy. Happily, the defense community has had to solve this problem for other places where the Sun does not shine, for example, to power submarines that have to be capable of staying at the bottom of the ocean for months on end.

Nuclear power plants similar to those used in submarines have provided the power sources for almost every long-distance space mission to date including those Voyager craft heading for the Oort clouds after a 40-year journey to the edge of the solar system. Adding a radioactive payload as a power source on a rocket is not risk-free but is not uncommon and the risks can be managed, and potential radiation can be minimized though the insurance costs implicit in any risk of accidentally irradiating America and adjacent continents could be prohibitive. However, think of it as a taking a bit of the Sun into space with you and it can seem like a relativity benign option.

Practically, it comes down to using radioactive isotopes to produce power from decay heat (thermoelectric generation) or from fission and fusion. The best option depends on the amount of power needed, the time scale over which it is required, the amount and type of gamma rays or X-ray or Y-ray ionizing radiation produced and the cost and complexity of containing that radiation.

However, consider that a uranium pellet encased in a grapefruit-sized ferrite core can be held in bare hands with no short-term or long-term material health impact. The power generated from these sources can be sufficient to produce temperatures high enough to split hydrogen and oxygen atoms from water. This offers the prospect of sustaining human life on any planet that has water but probably equally important provides the basics for manufacturing the liquid oxygen and hydrogen needed to return to Earth. This is all very exciting, but what we need is to understand is whether this is potentially useful for communications satellites in near Earth orbits. The start point is that any of these power sources produce far more power from a much smaller size than a solar panel array so potentially there are major weight and size savings that could be achieved.

Increasing the overall market for nuclear power generation by adding high-volume space applications would also help to reduce the cost of nuclear terrestrial energy providing a more space-efficient but equally carbon-friendly way of delivering power to the grid.

5.5.1 Thermoelectric Generation Using Radioisotope Power Sources for Communications Satellites?

Radio-isotopes have been used in space as a heat and power source for well over 50 years and are known as radio-isotope thermoelectric generators (RTGs). When used just to warm up electronic and mechanical components they are known as radio-isotope heating units (RHUs).

Plutonium, specifically Plutonium-238 (Pu-238) [11], has been widely used partly because it has been available as a by-product of the U.S. and Russian and other country weapon programs. It has a decay heat of –0.56W per gram and a half-life of 88 years. A typical RHU used to warm instruments to an efficient operational temperature would typically use just under 3 grams of plutonium in a box about 3 cm by 2.5 cm to produce 1W of power.

There are also many by-products of plutonium including Americium, produced when plutonium is bombarded with neutrons, for example, in a reactor or weapons test. Americium-241 is the most common flavor of Americium, manufactured from aging plutonium stocks. It is used in smoke detectors. Americium-241 has a half-life of 432 years but produces only 0.15W per gram (a quarter of the energy of plutonium). It produces higher levels of gamma radiation than plutonium and therefore requires more shielding (additional weight and cost). Note that shielding in manned missions is generally more onerous as it is generally considered inappropriate to irradiate astronauts at significantly elevated levels.

5.5.2 Production Costs for Americium and Plutonium

Because it is a by-product, Americium is significantly less expensive to manufacture. The cost of manufacturing a kilogram of plutonium has been estimated as $8 million. The European Space Agency is paying for AM-241 recovered from the United Kingdom's civil plutonium stocks [12] where this cost has essentially been amortized over many years of expensive nuclear power generation. The cost is therefore high but already paid for by the U.K. taxpayer.

For several decades, there has been enough plutonium available from civil and military nuclear programs including, for example, from the various nuclear missile reduction programs for space use either in its raw state or processed into AM-241. In 2011, NASA and the U.S. Department of Energy received $10 million of U.S. taxpayer funding to restart plutonium production with the intention of generating initially 1.5 kg per year at significantly lower cost [13]. Throughout the 1990s, the United States bought Pu-238 from Russia, in total about 16.5 kg, a by-product of START, the Strategic Arms Reduction Treaty [14] and Glasnost. A handful of Russian individuals became very wealthy from these exchanges. When President Putin came to power, Russia decided it would no longer be a source of supply, hence the focus on U.S.-based production capability.

Plutonium is produced by irradiating Neptunium-237, a radioisotope with a half-life of just over 2 million years.

5.5.3 How Long Do Radio-Isotope Thermoelectric Generators Last?

The Voyager spacecraft that have just left the solar system are expected to keep sending back signals to Earth until 2025, the best part of 50 years of operational life.

Voyager 1 is now over 20 billion km from Earth, more than 139 times the distance from the Earth to the Sun. Voyager 2 is 11 billion miles away. In December 2017, Voyager 1 used its trajectory maneuver thrusters for the first time in 37 years [15]. This was only possible with a spacecraft with onboard, long-term electrical power.

There are several dozen RTGs presently powering U.S. and Russian space vehicles. Cassini, for example, sent to explore Saturn's rings, was powered by three RTGs providing 870W of power from 33 kg of plutonium oxide. As you may remember, there was a planned deorbit into Saturn's atmosphere on September 15, 2017 [16]. The Pathfinder Mars robot lander launched in 1996 had three RTGs each with 2.7 grams of plutonium-238 oxide producing 35W of power and 1W of heat.

The state-of-the-art RTGs today are known as general-purpose heat source (GPHS) modules [17]. The latest Mars Rover, Curiosity, had (at the time of this writing) traveled 18 km across the surface of Mars powered by 8 GPHS units containing a total of 4.8 kg of plutonium oxide producing 2 kW of thermal power generating 110W of electricity. The Mars Rover has an Earth weight of 890 kg, significantly heavier than a Caterham sports car.

The New Horizons spacecraft that flew by Pluto in July 2015 was launched in 2006. The 250-W, 30-V RTG produced 200W from 10.9 kg of Pu-238 oxide, which had reduced to 200W by the time the craft arrived near Pluto. The vehicle has 65 kg of hydrazine available to control 16 Aerojet thrusters generating a few newtons of power.

Russian RTGs are apparently still operational in orbit on Cosmos navigation satellites launched in 1965. China's lunar lander apparently uses Pu-238-based RTGs.

5.5.4 Heat-to-Electric Conversion Using Stirling Radioisotope Generators

RTGs turn heat into energy by using simple thermocouples [18]. These are almost completely reliable (no known or recorded in service failures) but not efficient (2 kW to heat to produce 10W of electricity; see above, although the extra heat can also be useful).The alternative is to use a Stirling engine.

Stirling Engines [19] can produce at least 4 times more electricity from a gram of plutonium when compared to a simple thermocouple. There is a hot end which could be, for example, at 650°C which heats up helium which then drives a free piston reciprocating in a linear alternator powered by the temperature difference either side of the piston. Two SRGs working on about 500W of thermal power should produce about 140W of electric power from a kilogram of Pu-238.

Invented by Robert Stirling in 1817, the Stirling Engine [19] is being promoted as a semimagic way of turning waste heat from domestic and industrial processes into useful electricity. Our interest in the context of space is its capability to scale to high-temperature gradients. Although not as reliable as thermocouples, a space-qualified SRG is not intrinsically unreliable and several small engines coupled together will have a high level of redundancy.

The Idaho National Laboratory's (INL) Center for Space Nuclear Research (CSNR) [20] in collaboration with NASA is developing an RTG-powered hopper vehicle for Mars exploration, supported by NASA. When stationary, the RTG breathes in carbon dioxide from the Martian atmosphere, compresses it through a Stirling engine, and freezes it. A beryllium core stores heat energy to fuel an explosive vaporization for the next hop. When ready for the next hop, nuclear heat vaporizes the carbon dioxide, creating a jet capable of propelling the craft to an altitude of 1,000m and a hop of 15 km with payloads of up to 200 kg. The surface gravity on Mars is only 38% of the surface gravity on Earth. If you weigh 100 kg on Earth, you will only weigh 38 kg on Mars, an easy way to lose weight.

5.6 Fission and Fusion

Not content with radioisotope thermoelectric generators, Russia has invested significant development in fission reactors for space power systems. Just a reminder, fission and fusion are both nuclear reactions that produce energy, but fission does it by splitting a heavy, unstable nucleus into two lighter nuclei, and fusion crashes two light nuclei combine to release a vast amount of energy very quickly [21]. Fission is recreating the Sun in a small package; fusion is capturing the power of the Big Bang, an altogether more cataclysmic process. Russia has used over 30 fission reactors in space; the United States has flown only one, the System for Nuclear Auxiliary Power in 1965.

From 1959 to 1973, there was a U.S. nuclear rocket program, Nuclear Engine for Rocket Vehicle Applications (NERVA), working on using nuclear power rather than chemical power for the latter stages of launches. NERVA used graphite-core reactors heating hydrogen and expelling it through a nozzle. Some 20 engines were tested in Nevada and yielded thrust up to more than half that of the space shuttle launchers. Generally, it was felt that this would be altogether too hazardous for Earth-bound mortals and the focus shifted to propulsion in space. A $19 million contract has been placed by NASA with specialist nuclear energy company, BWXT Nuclear Energy Incorporated, to study the feasibility of a nuclear thermal rocket [22].

In 1958, the U.S. Project Orion planned to launch a 1,000-ton spacecraft using a series of nuclear explosions. The project was stopped in 1958 by General Atomics when the Atmospheric Test Ban Treaty made it illegal. However, Russia pressed on with fission reactors for space using uranium carbide fuel at high temperature.

5.7 Why Uranium Is Cheaper Than Plutonium

Uranium is cheaper than plutonium because you can dig it out of the ground. Well at least, you can dig Uranium 235 out of the ground and then refine it into something more useful in terms of realizable energy content (for example, Uranium 233).

There are three major fissile isotopes, Uranium 235, used in the Hiroshima bomb and most nuclear power reactors, Uranium 233 used in Thorium reactors but not in weapons and Plutonium 238. It is a bit like diesel and petrol and paraffin; you take a basic ingredient and transmute it into something else, in the case of fissile isotopes, by firing neutrons at whatever you happen to have available.

The first nuclear weapons used uranium because plutonium had to be manufactured by neutron bombardment and to make plutonium you need a lot of neutrons and the only realistic way of getting these is a uranium-based fission reaction [23]. Plutonium has more energy density than uranium and it is easier to get plutonium to a critical mass than uranium. For weapons systems, this is an advantage but for a communications satellite energy source, a disadvantage. Plutonium is also problematic in terms of the damage it can cause to humans. Plutonium produces lots of alpha radiation rather than beta or gamma radiation.

Of the three types of ionizing radiation, alpha is the least penetrating while gamma is the most penetrating. However, plutonium gets into humans via the bloodstream via the lungs then keeps going into our bones, liver, and all other vital organs where it can stay for decades before it kills us, although if the dose is big enough, it can kill us alarmingly quickly as firefighters at Chernobyl and Fukushima tragically demonstrated. When alpha rays get into our cells, they cause between 10 and 1,000 times more chromosomal damage than beta or gamma rays. Polonium, also a by-product of uranium, has similarly devastating medical effects particularly when added to tea [24].

Whatever the source, there will be a need to protect electronics and for manned missions, there will be a need to protect humans from a mix of radiation products. Shielding is dependent on the mission or application. Lithium hydride in stainless steel cans, for example, is often used for neutron shielding.

Another consideration is the time scale over which these sources are active. The half-life of Pu-239 is 24,100 years. Radioactive contaminants are dangerous for 10 to 20 times the length of their half-lives, meaning that dangerous plutonium released to the environment today will be with us for half a million years, which make this a depressingly long-term problem.

5.8 Back to Russia and the United States and China

In 2010, the Russian Presidential Commission on Modernization and Technology Development of Russia's Economy [25] allocated funds to design a megawatt nuclear power propulsion unit (NPPU) for long-haul interplanetary missions.

This indirectly prompted the United States and China and other nuclear states to review their own research programs. The United States had been working on conversion systems that could efficiently translate high temperatures from fission processes into electricity using heat pipes to transfer energy from the reactor core [26] or Stirling or Brayton cycle converters [27]. Heat pipes are essentially high-tech kettles, exploiting energy release from changes of state [28].

The Brayton cycle convertors, if you have the energy and enthusiasm to follow the URL links, are essentially based on materials innovation but are basically a kettle using carbon dioxide to make the perfect cup of tea (without added polonium) [29]. The World Nuclear Association [30] provides a thorough summary of progress over the last 30 years with compact fission reactors for space applications. As a summary, the heat is taken from the fissile core fuel pins to heat pipes filled with sodium vapor, which transfers the heat to heat exchangers to heat up a gas that is usually a mixture of helium and xenon. The hot gas is then used to power a Stirling or Brayton engine. These devices are capable of producing many kilowatts of continuous power for very long periods of time with ongoing research on nuclear electric propulsion systems driven by plasma with a power of the order of 100 kW. These nuclear-propelled and nuclear-powered space vehicles should provide a faster, more comfortable way of getting to Mars and beyond and provide a convenient excuse for maintaining a nuclear development program with possible space weaponry application.

Ion engines powered by small nuclear reactors are theoretically capable of producing 20 kW or more of propulsion power over a 7 to 10-year lifetime with high fuel efficiency. There are also plans to produce megawatt power sources but the reactors weigh between 30 and 40 tons.

The French ERATO program was based on combining three 20-kW turboelectric power systems all using a Brayton cycle converter with helium-xenon as working fluid. The first system was a sodium-cooled UO2-fueled fast reactor operating at 670°C, the second was a high-temperature gas-cooled reactor (thermal or epithermal neutron spectrum) working at 840°C, the third was a lithium-cooled UN-fueled fast reactor working at 1,150°C. Thermal neutrons are neutrons in thermal equilibrium with a surrounding medium. Epithermal neutrons have a kinetic energy greater than thermal. Epithermal neutrons produce higher core efficiency.

5.9 Regulatory Issues of Launching Radioactive Material into Space

The regulatory issues associated with nuclear powered satellites are dealt with by the Office for Outer Space Affairs (UNOOSA) [31] under the administration of the United Nations. UNOOSA implements policy decisions taken by the Committee on Peaceful Uses of Outer Space (COPUOS) [32] set up in 1959 and now supported by 75 member states.

5.10 Risks Associated with Launching Radioactive Material into Space

Environmental groups are not always happy at the prospect of firing small or large amounts of radioactive material into space. When the Cassini-Huygens probe was launched in 1997, the U.S. Department of Energy estimated the chances of a launch accident that would release radiation into the atmosphere at 1 in 350. It was estimated that a worst-case scenario of total dispersal of onboard plutonium would spread the equivalent radiation of 80% of the average annual dosage in North America from background radiation over an area with a radius of 105 km though the methodologies used in these calculations are always open to interpretation and legal challenge.

It would be different if uranium and plutonium power sources could be produced on the Moon and then shipped back to near-Earth orbit, a not altogether impossible prospect in a 30 to 50-year time frame.

5.11 Uranium in the News

Speaking of which, uranium has recently been in the news [33]. The LIGO [34] and VIRGO [35] gravitational wave detectors at the time of this writing had detected gravitational wave energy generated by the merging of two neutron stars. This followed the first detection of a gravitational wave from two collapsing black holes. These events are now calculated to happen every 15 minutes somewhere in the universe. Merging neutron stars are a source of the heaviest chemical elements on Earth including uranium, platinum, and gold ejected as a fireball of radioactive chemical elements known as a kilonova, accompanied by a burst of gamma rays and visible light, which were detected by a combination of Earth and sky-based telescopes (NASA Fermi and ESA Integral). The bursts were detected 2 seconds after the gravitational wave. It all happened 130 million light-years away in the constellation Hydra. Neutron stars are the remains of large stars whose cores have collapsed producing a tiny ball of immensely dense neutrons. A thimble full of neutron star is the equivalent of a small mountain in weight. Two neutron stars colliding either produce a single larger neutron star or, depending on their temperature, spin speed, and mass, a black hole. The gamma ray burst and flash of light indicate that this latest measurement was a merging of two neutron stars (gamma rays and light rays would not normally escape from a black hole). The events also have a different wave signature. Merging black holes produce a wave that is observable for a fraction of a second. When two neutron stars merge, the gravitational waves are observed for about a minute.

The mysteries of gravitational waves, first predicted in 1916 by Albert Einstein, may see remote to the present-day reality of 5G and satellites, but these discoveries mark a significant advance in our understanding of energy and its nuclear origin and radiation characteristics.

5.12 Radiation in Space: Photons or Neutrons, the Final Choice?

The 2011/2012 space mission to Mars measured this radiation from all sources during the 36 weeks that it took to get to Mars. The spacecraft was exposed to an average of 1.8 mSv per day, suggesting a total exposure of 660 mSv for astronauts and their instruments on a two-way trip. The equivalent radiation dose for astronauts on the International Space Station is of the order of 100 mSv over 6 months. Radiation exposure is therefore a significant motivation for the building of faster rockets.

Radiation can also cause hardware failure. First-generation Globalstar satellites, for example, suffered failures with the onboard RF power amplifiers, and

Boeing had systematic failures in orbit with spacecraft using their 702 bus [36]. Hardware damage from radiation is a well-understood phenomenon with well-established mitigation measures.

The Boeing issues were related to fogging on the solar power concentrator, which reduced output power from 18 kW to 12 kW with litigation threatened by customers including PanAmSat, Thuraya, XM Satellite Radio, and Telesat. The insurance underwriters are also pursuing compensation based a claim of systematic system failure.

Solar panels are vulnerable to space damage and have to be protected with expensive space qualified glass. Nuclear power sources are arguably significantly more reliable, certainly longer-lasting with the potential to scale to the tens of kilowatts needed for next-generation mobile and fixed broadband satellite systems.

5.13 CubeSat Innovation

The Boeing satellites are very large geostationary satellites, but innovation is also being applied to very small satellites including CubeSats.

This includes CubeSats with optical transceivers in which the laser is hard-mounted to the spacecraft body with the orientation of the CubeSat determining the direction of the beam.

The miniature satellites are 10 cm × 10 cm × 10 cm (4 inch cubes) and their intended use is for high-speed intersatellite and satellite-to-Earth communications or to test novel propulsion systems including systems that use water as a propellant.

The satellites will also test control systems including autonomous docking capabilities with other CubeSats using low-cost sensors or docking with larger satellites [37].

The ability to point accurately is critical to the throughput of an optical transceiver. Throughputs of 200 Mbps are claimed to be achievable in free space.

5.14 Quantum Computing Using Optical Space-Based Transceivers

Japan's National Research and Development Agency (NICT) [38] have developed what they claim is the world's smallest and lightest quantum communication transmitter onboard the microsatellite SOCRATES. The satellite weighs 6 kg and is 17.8 cm in length, 11.4 cm wide, and 26.8 cm high. The satellite transmits a laser signal to Earth at a rate of 10 million bits per second from an altitude of 600 km at a speed of 7 km per second. The project is targeted at producing an ultrasecure communications network [39].

5.15 Smartphones in Space: A Megawatt, Very Mobile Network

I am not aware of specific proposals to use CubeSats as part of a mobile and fixed broadband network. The general assumption is that they would have insufficient power budget and or antenna aperture/antenna gain to support higher bandwidth space to Earth and Earth-to-space communication and are better suited to short and occasional but periodic transmission bursts from IoT devices. However, smartphones in terrestrial networks can receive and send data to and from multiple base stations [40] and a similar approach could be taken with very high-count CubeSat constellations. Sending 1 million ruggedized smartphones to the Karman limit with 1W of output power from each device would produce a 1-MW diversity transmit downlink with sufficient device density to deliver substantial diversity gain.

This is not as fanciful as it might seem. In 2013, a NASA sponsored team [41] launched three Phonesat satellites into space based on a consumer-grade smartphone. This was motivated by the recognition that the processing power in an average smartphone coupled to a 40-megapixel camera and sophisticated battery with even more sophisticated power management was equivalent and often better than many small satellites but at a cost several orders of magnitude lower. Most of the team then left to start a company focused on building satellites from low-cost, off-the-shelf commercial components coupled to an imaging and Earth observation database [42].

5.16 Other Power Sources in Space

NASA has also been working on other power sources in space including closed-cycle proton exchange membrane fuel cells (PEMFC) with outputs of between 1 and 10 kW, scalable up to 100 kW with energy weight ratios of the order of 250W to 350W per kilogram of cell weight and 10,000-hour service life [43]. PEMFCs are electromechanical power generation devices that convert hydrogen and oxygen reactants into electrical power, heat, and water. The hydrogen and oxygen can be shared with propulsion systems and the water by-product can either be used by humans or potentially used as a jet thruster for pitch and pointing control for satellites. They provide a useful alternative to battery storage including applications where the solar panels are not receiving solar power for significant periods due to orbit trajectory or pitch and pointing requirements. These power sources presently have a relatively limited service life expectation of between 1 and 2 years constrained by membrane performance (see Table 5.2).

5.17 Satellites, Energy Efficiency, and Carbon Footprint

Later in the book, we discuss some of the emerging issues of energy costs in dense terrestrial 5G networks. Although LEO, MEO, and GSO satellite constellations are not specifically being targeted at improving the overall energy efficiency of terrestrial networks, it could be argued that they have contributions to make in several areas including energy efficient backhaul. Because it is sunnier in space and more solar power is available, energy costs should be lower. Satellites could also help improve the carbon footprint of 5G terrestrial networks.

5.18 Antenna Innovation

Finally, the delivery economics of LEO, MEO, and GSO satellite constellations are being transformed by antenna technology innovations both on the satellite and on Earth-based user terminals and IoT devices. We are going to cover this in Chapter 6, but essentially the story can be summarized as isotropic gain, the art of ensuring that RF energy gets sent in the right direction combined with energy rejection, the ability to null out unwanted signal energy.

5.19 5G and Satellite: The Nuclear Option

The relevance of nuclear power sources to modern communications systems may not be immediately obvious, but for deep space communication where the Sun does not shine there are no other available options.

Table 5.2
Power Source Comparisons: Photons versus Neutrons versus Fuel Cells

GSO Solar Array	Radioisotope Thermoelectric Generators	Stirling or Brayton Cycle Engines	Proton Exchange Membrane Fuel Cells
30-m span solar panels produce 15 kW reducing to 12 kW at end of life (15 years)	Milliwatts to watts to kilowatts (general-purpose heat source modules), simple thermocouple, no moving parts, 100% reliable, 50-year life	4 times better conversion efficiency than RTGs, 500W of thermal power = 140W of electric power from 1 kg of Pu-238 oxide, 15-year life?	Efficient nonradioactive option, high-energy weight ratio, can be shared with propulsion system, liquid hydrogen and liquid oxygen, water as a by-product, 10,000-hour service life (1 to 2 years)

The two Voyager spacecraft that have just left the solar system after 40 years are on their way to the Oort clouds, which they will reach in 300 years' time. It will be 30,000 years before they emerge from the other side with still many thousands of years to go before the next galaxy appears on the horizon.

The communication system will carry on working until at least 2025, which means that the Voyager transceivers will have been operating for nearly 50 years.

Mr. Musk's mission to Mars will require a range of isotope-based and fissile power systems for propulsion, onboard power, and hydrogen and oxygen production to sustain life on Mars and the production of liquid fuel for the return to Earth. NASA, China, and Russia are all working on new generations of small nuclear reactors and isotope power sources.

Mr. Musk's very large rocket can either take a relatively small payload (a few astronauts and their baggage allowance) to Mars or a very large payload, potentially several dozen satellites per launch into near LEO and it can be assumed that this will be the vehicle that takes the 4,000 LEO satellites into space at a cost base several orders of magnitude below present satellite systems. OneWeb and LeoSat and OneWeb have similar plans for high-count LEO constellations and the required cash courtesy of Mr. Bezos and his new rocket company (Blue Origin) and Mr. Branson (Virgin Galactic).

5.20 Summary

This chapter has drawn parallels with the previous chapter on rockets. In particular, we have argued that deep space exploration has required innovation in propulsion and power technologies that can be equally applied to rockets and the payloads that they carry.

In practical terms, this means that near-Earth orbiting satellites including LEO, MEO, and GSO constellations now have a much wider choice of propulsion and power systems. Examples include the new generation of electric satellites that can either sail into deep space or sail themselves into a near-space orbit with station-keeping and then be managed from the solar power budget rather than space and weight-limited hydrazine fuel sources.

The use of nuclear power sources is common place in deep space missions and unavoidable for missions beyond Mars where the Sun shines progressively more weakly.

Mars missions that are planned both by the private sector and sovereign nation space programs are focusing significant attention on a new generation of radioisotope and fissile radioactive power sources that have energy to weight and size ratios several orders of magnitude greater than any other nonnuclear power source.

While the economics of near-Earth communication do not presently support the widespread use of these alternative power systems, there may well come a point where the systems used for deep space exploration are repurposed to power LEO, MEO, and GSO communication satellites. Avoiding the need for solar panel arrays improves the pointing accuracy of satellites and their maneuverability. This may be an elegant option for optimizing progressive pitch as LEO satellites move towards and away from the equator. This, in turn, should help to optimize the angular power separation of LEO signal energy from MEO and GSO space and terrestrial-based receivers (and potentially 5G terrestrial receivers as well).

The possibility that this will be the only way to meet the required protection ratios for LEO, MEO, and GSO coexistence in Ku-band, K-band, and Ka-band may be a compelling argument for the nuclear option, although the associated cost and risk needs to be precisely assessed. There may be lower-risk, lower-cost options such as fuel cells that may emerge as a credible alternative.

Generating sufficient power cost-effectively and energy efficiently is critical for terrestrial and space-based networks. Sending that power in the wrong direction makes no sense at all.

References

[1] http://www.azurspace.com/images/products/0004301-00-01_DB_3G30W-Advanced.pdf.

[2] http://www.spectrolab.com/history.htm.

[3] https://www.nasa.gov/centers/glenn/about/fs21grc.html.

[4] http://www.rocket.com/propulsion-systems/electric-propulsion.

[5] http://www.unitconversion.org/power/kilowatts-to-newton-meters-per-second-conversion.html.

[6] http://hyperphysics.phy-astr.gsu.edu/hbase/magnetic/Hall.html.

[7] https://www.safran-aircraft-engines.com/media/boeing-selects-safran-ppsr5000-electric-thruster-use-commercial-satellite-20170627.

[8] http://www.esa.int/Our_Activities/Telecommunications_Integrated_Applications/Partner_Programme_Electra.

[9] http://www.ohb-sweden.se/space-missions/.

[10] http://spacenews.com/european-soyuz-launches-smallgeo-satellite-for-hispasat/.

[11] https://solarsystem.nasa.gov/rps/docs/APP%20RPS%20Pu-238%20FS%2012-10-12.pdf

[12] http://www.world-nuclear-news.org/F-Cosmic-recycling-2107141.html.

[13] http://ne.oregonstate.edu/rebuilding-supply-pu-238.

[14] http://www.nti.org/learn/treaties-and-regimes/treaties-between-united-states-america-and-union-soviet-socialist-republics-strategic-offensive-reductions-start-i-start-ii/.

[15] https://www.space.com/38967-voyager-1-fires-backup-thrusters-after-37-years.html.

[16] https://saturn.jpl.nasa.gov/mission/saturn-tour/where-is-cassini-now/.

[17] https://solarsystem.nasa.gov/rps/types.cfm.

[18] https://solarsystem.nasa.gov/rps/rtg.cfm.

[19] https://www.stirlingengine.com/faq/.

[20] http://csnr.usra.edu/public/default.cfm?content=347.

[21] https://nuclear.duke-energy.com/2013/01/30/fission-vs-fusion-whats-the-difference.

[22] https://phys.org/news/2017-08-nasa-reignites-nuclear-thermal-rockets.html.

[23] https://www.reddit.com/r/explainlikeimfive/comments/1idbux/eli5_whats_the_difference_between_uranium_and/.

[24] https://www.medicalnewstoday.com/articles/58088.php.

[25] http://en.kremlin.ru/events/president/news/10453.

[26] http://propagation.ece.gatech.edu/ECE6390/project/Fall2010/Projects/group5/sites.google.com/site/inasa6390/design-overview/Power/heatpipe-power-system.html.

[27] http://www.adl.gatech.edu/research/spg/papers/InCA_IECR2012paper.pdf.

[28] https://www.nasa.gov/mission_pages/station/research/news/heat_pipes.html.

[29] http://energy.sandia.gov/energy/renewable-energy/supercritical-co2/.

[30] http://www.world-nuclear.org/information-library/non-power-nuclear-applications/transport/nuclear-reactors-for-space.aspx.

[31] http://www.unoosa.org/

[32] http://www.unoosa.org/oosa/en/ourwork/copuos/2017/index.html

[33] https://www.livescience.com/60695-why-gravitational-wave-discovery-matters.html?utm_source=notification.

[34] https://www.ligo.caltech.edu/.

[35] http://www.virgo-gw.eu/.

[36] https://www.flightglobal.com/news/articles/insurers-may-pursue-boeing-over-satellite-power-fail-171593/.

[37] https://www.nasa.gov/press-release/cubesat-to-demonstrate-miniature-laser-communications-in-orbit.

[38] https://www.nict.go.jp/en/about/.

[39] https://phys.org/news/2017-07-world-space-quantum-microsatellite.html#jCp.

[40] http://www.networkcomputing.com/wireless/lte-broadcast-horizon/1433776652.

[41] https://www.nasa.gov/offices/oct/feature/how-a-nasa-team-turned-a-smartphone-into-a-satellite-business.

[42] https://www.planet.com/.

[43] https://www.nasa.gov/content/space-applications-of-hydrogen-and-fuel-cells.

6

Antenna Innovation

6.1 The Impact of Antenna Innovation on Energy Costs in Terrestrial and Nonterrestrial Networks

6.1.1 The Function of Antennas in Noise Limited Networks

In Chapters 1 through 5, we touched on the importance of energy efficiency in terrestrial and nonterrestrial networks. In terrestrial networks, energy efficiency is directly related to energy cost and therefore has a direct impact on network operational costs. These costs vary from country to country and can be particularly problematic in countries with a limited electricity grid, for example, in parts of Africa where the only power available is either solar or diesel. Solar panels disappear from remote sites and supplying diesel incurs additional operational overheads. Solar panels also require battery backup. Lead acid and lithium-based batteries are expensive, take up space, and have limited capacity and a limited life.

Energy costs in terrestrial networks are a composite of the RF power needed across the radio interface, the baseband processing overhead, and the backhaul overhead. One might think that as network density increases, energy costs would reduce as less RF power is needed to service local users and devices. In practice, the opposite is true partly because RF interference becomes a dominant constraint and partly due to the power required for backhaul. The extra energy needed to support denser networks is a subject of debate but one reliable estimate suggested that energy costs could multiply by a factor of three as networks transition from kilometer cell sizes to cell radii of 100m or less, according to informal discussions with vendors. The good news is that LTE is more power-efficient than 3G despite modulation and multiplexing that requires more linear amplification. This is due to the need for 3G to deliver symbols at equal power level, which is easily compromised by inaccuracies in uplink and downlink power control loops. 4G LTE and 5G requires users to be received at a base station at the same time (realized by using timing advance and using the cyclic prefix as a time-domain guard band), but they can be at different power levels, avoiding the overhead of complex band-hungry power control.

In space, it could be argued that as with solar-powered terrestrial base stations, the energy comes for free, but in practice there are associated costs. Theft is not a problem in space, but the size, weight, and build quality of the solar panels on a satellite add to the cost of the satellite and increases launch cost. Antenna arrays can be damaged by debris impact and degrade over the lifetime of the satellite. In Chapter 5, we also pointed out that large solar panel arrays made satellites less maneuverable. This can be a problem for satellites that implement progressive pitch control partly due to the additional spin mass, but the solar panels also ideally need to point towards the Sun for as much time as possible.

As a rule of thumb, about half the power requirements in a terrestrial base station or access points or Wi-Fi transponder are related to the RF power budget which in turn is related to the link budget (see Chapter 2). However, if a transceiver is working close to its receive sensitivity or maximum power limit, then additional channel coding will be incurred. This absorbs radio layer capacity but also consumes additional clock cycles which increase power consumption.

It is therefore important to ensure that RF energy is sent where it is needed. Ideally, antennas would produce a narrow beam of concentrated energy, effectively recreating the characteristics of guided media such as copper, cable, or fiber.

They achieve this through isotropic gain. Narrow beam antennas include fractional beam width antennas defined as antennas with a 3-dB half-power beamwidth of between 0.5° and 1.5°. These can deliver of the order of 40 to 50 dBi of isotropic gain.

However, narrow beam and fractional beam antennas have a cost in terms of the aperture size, cost, and weight of the antenna, particularly at lower frequencies/longer wavelengths. If there is a problem with pointing of the antenna, for example, in terrestrial systems due to high winds or in satellites due to poor yaw and pitch control, then much of this isotropic gain will be absorbed by pointing loss.

There are antennas at both ends of the radio link. Generally, terrestrial base stations, terrestrial access points, and satellites have sufficient space to support high-performance, high-gain antennas. This includes antennas systems that can adapt to changing noise conditions, for example, high levels of noise coming from a particular angle of arrival.

This is harder to achieve in small form factor user and IoT devices, particularly at lower frequencies and longer wavelengths where space constraints mean that antennas are not inherently efficient with less than optimum ground planes. In smartphones, this is made worse by the need to support multiple antennas in a small space. It is not uncommon to have user and IoT devices working at sub-1 GHz with a negative gain of the order of −7 to −10 dB.

Antennas are happiest working within 10% of their center frequency. They can be forced to work over wider bandwidths either by switching in additional lengths of antenna or by electrically lengthening the antenna, but this will compromise the noise matching and power matching of the antenna. The physics of this process are outside the scope of this book, but can be researched by delving into the inner magic of the Smith Chart developed by Mr. Smith in 1939 [1].

Antennas working across wider bandwidths will also be vulnerable to hand capacitance effects where how you hold the phone has a major impact on the RF performance of the device. There are adaptive matching techniques to mitigate this, but these, in turn, have a power budget cost.

It is hard to realize useful directivity in small form factor user and IoT devices. In base stations and Wi-Fi access points, narrow beam antennas should reduce the amount of unwanted energy transferred into spectrally and geographically adjacent radio systems, although not if they are pointing in the wrong direction.

Satellites are essentially base stations in the sky normally using dish antennas to focus on specific geographic areas with the objective of providing enough flux density for ground-based receivers to detect a wanted signal above the noise on the radio channel, for example, to receive TV broadcasts.

For two-way communications, there has to be sufficient gain on the satellite receive antenna to overcome the uplink path loss, bearing in mind that the device on the ground may have relatively low output power of the order of 1W or 2W. These are usually described as spot beam antennas.

If implemented using dish antennas, the spot beams can be mechanically pointed to provide coverage and capacity on demand. If implemented using flat panel antennas with multiple antenna elements, the beam forming is achieved electrically by changing the phase of each antenna element.

At higher frequencies and shorter wavelengths, these antenna systems provide highly focused coverage. An example is the V-band low Earth orbit (LEO) constellation proposed by Boeing at 37.5–40 GHz and 51.4–52.4 GHz with 1-GHz channels supporting cells with a diameter of between 9 and 11 km (Figure 6.1). It is proposed that the satellites would also have C-band antennas.

Self-evidently, antennas on ground devices need to be capable of looking at the sky either physically or electrically. This results in some distinctive

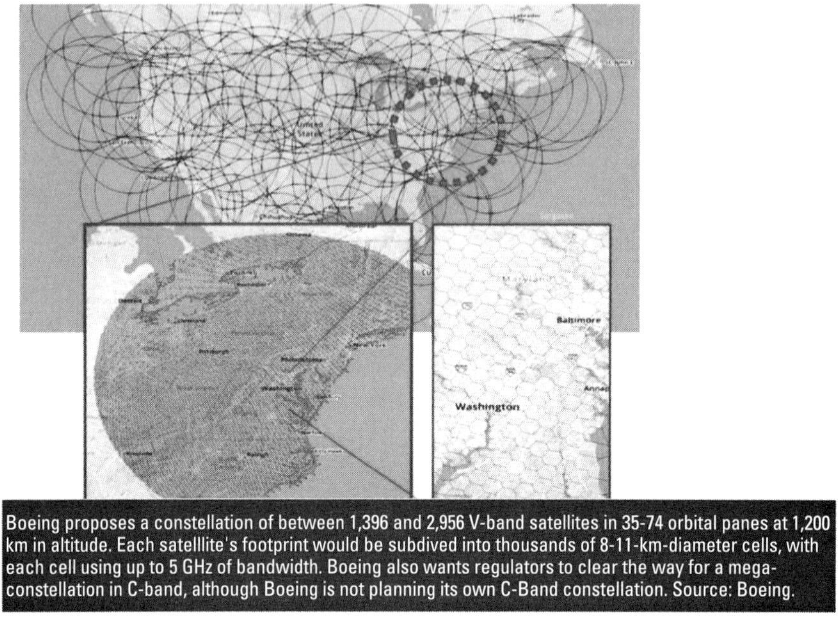

Figure 6.1 Boeing LEO constellation showing terrestrial cell patterns. (Thanks to the Boeing Corporation.)

antenna designs, for example, the Iridium user terminal in Figure 6.2 with the L-band antenna.

6.1.2 The Function of Antennas in Interference Limited Networks and Satellite and Terrestrial Coexistence

This brings us to the function of antennas in interference limited networks. If two or more simple collinear antennas (long single-pole antennas containing an E and H plane) are moved close to each other (less than a fraction of a wavelength apart), then they will start coupling together and the phase of each antenna will be influenced to create a change in the gain and null of the radiation pattern from the combined antennas. This means that interference from a particular direction can be nulled out. This technique has been used for over 50 years in very high frequency (VHF) and ultrahigh frequency (UHF) networks, for example, to protect emergency service radio systems from unwanted TV signal energy.

Modern antenna arrays achieve the same effect by electrically changing the phase relationship between antenna elements. This has the significant advantage that the radiation pattern can be changed in response to changing interference conditions, for example, high levels of unwanted noise though more commonly high levels of unwanted signal energy (interference).

Figure 6.2 Iridium user terminals with L-band antennas and Wi-Fi unit for local connectivity.

The beam pattern can be changed in azimuth to minimize interference coming sideways from left or right of an antenna or in elevation. As covered in Chapter 2, satellites can be at low elevation. For example, satellite TV dishes in high northern and southern latitudes are pointing close to the horizon in order to receive signals from satellites broadcasting TV signals from geostationary orbits over the equator. The same dishes at the equator, for example, in Singapore, are more likely to be pointing directly upwards provided there is a GSO satellite directly overhead.

LEO and MEO constellations can be anything from low elevation to directly overhead. Generally, the best link budget will be directly upwards as this minimizes the amount of atmosphere through which the signal has to travel, but this requires a high-count satellite constellation.

However, it can be seen that there are substantial opportunities for achieving angular power separation between terrestrial and nonterrestrial networks. By implication, this makes cosharing of satellite spectrum with terrestrial networks feasible and potentially commercially attractive. As we shall see in later chapters, this is a contentious issue and open to technical and legal challenge, but the spectral efficiency gains from frequency reuse could be substantial.

We revisit angular power separation in Chapter 7, but before we get there, we should usefully review some of the ways in which terrestrial and satellite antennas and antenna arrays need to be matched to specific channel conditions.

6.1.3 Four Things Antennas Are Supposed to Do but Cannot Do at the Same Time

Figure 6.3 summarizes the four functions that terrestrial antennas can perform: spatial diversity, coherent gain, interference mitigation, and spatial multiplexing. Each function requires specific baseband processing, so only one of these functions can be performed at any one time.

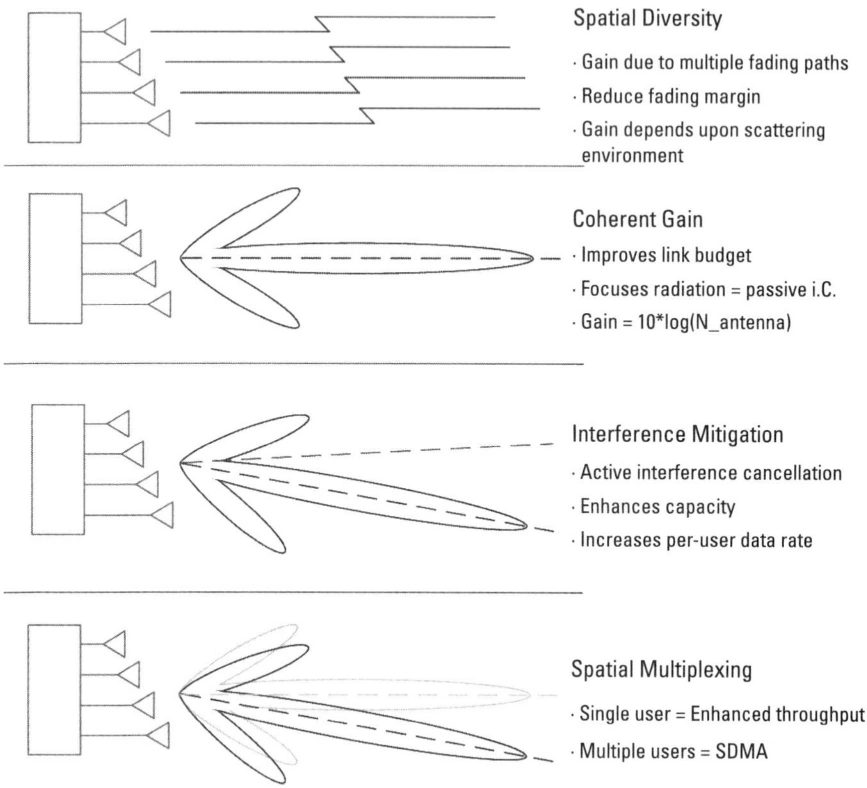

Figure 6.3 Four things antennas are supposed to do but cannot do at the same time. (Thanks to Arraycomm.)

6.1.3.1 Spatial Diversity

In terrestrial networks, both in-building and in urban and rural environments, there can be significant signal energy that arrives at a receiver having bounced off hard surfaces on the way and the composite signal transmitted will have followed several different paths on its journey to the receiver. This is known as scattering. Spatial diversity is the use of antennas to capture each of these paths so they combine constructively in the receiver. The signals need to be aligned in time, which is achieved by a channel equalizer and in phase, achieved by the use of a phase locked loop. This reduces the required fading margin, although the gain achieved is dependent on the number of signal paths and their relative strength.

At higher frequencies and shorter wavelengths, particularly in the millimeter band above 30 GHz, any surface roughness on walls or other reflective surfaces will be similar to the wavelengths of the radio signal being reflected and will result in significant signal absorption. This is why spatial diversity using multipath becomes less effective as wavelengths get shorter.

6.1.3.2 Coherent Gain

Coherent gain is where several antennas are used to collect the same signal following the same path from transmitter to receiver. Coherent gain is most useful in line-of-sight conditions, for example, from nearly always nearly overhead high-count LEO constellations.

6.1.3.3 Interference Mitigation

As covered briefly already, this is where phase offsets between multiple antennas or multiple antenna elements are changed in order to null out unwanted signal energy.

6.1.3.4 Spatial Multiplexing

This is widely used in TDD Wi-Fi and is also intensively standardized in 5G standards as a mechanism for achieving very high data rates within a small area, often indoors. User bits are mapped on to symbols, which are then coded on to multiple antennas or antenna elements, which effectively create a deterministic multipath that can be correlated with a similar number of antennas and antenna elements at the receiver. They are more effective in TDD systems because the uplink and downlink are on the same frequency and the channels are therefore reciprocal. Spatial multiplexing does not scale efficiently to larger cells and higher frequencies. In many propagation environments, FDD will provide a higher throughput gain. Separating the receive path from the transmit path in the frequency domain (frequency duplex separation) in a user or IoT device delivers a sensitivity gain. FDD also provides frequency separation between user and IoT devices and access points. An example is a home with multiple Wi-Fi access points deployed to support an Amazon Echo network or Google Home. This is the reason why the latest 802.11ax standard supports FDD.

6.2 Signals from Multiple Access Points, Multiple Base Stations, and/or Multiple Satellites

An additional option is to send the same signal from multiple access points and or multiple base stations and from multiple satellites. This is done in LTE broadcast [2] to deliver a link budget gain by summing signals from multiple sources and is one of the mechanisms proposed for CubeSats to sum small, low-power transmitters together to provide effectively one very large aperture (horizon-to-horizon) antenna.

6.3 Satellite Channel Models and Antennas: Standards as a Starting Point

Hopefully, it is clear that the choice of antenna system is determined by the characteristics of the radio channel, which, in turn, is determined by channel models that are derived from measurements and empirical observation (see also Chapter 2).

Earth-to-space and space-to-Earth propagation has been intensively studied as a by-product of near-Earth and deep space communication but only in the context of existing space network GSO MEO, and LEO topologies. Modeling high-count constellations is less advanced partly because these constellations do not exist yet and therefore do not have the empirical data available to calibrate existing or future theoretical models.

In October 2017 this was addressed by the 3GPP nonterrestrial networks group sponsored by Thales Alenias, Dish Networks/Echostar and Hughes Networks, Inmarsat, and Ligado [3].

There have been several unsuccessful attempts to develop integrated mobile broadband and satellite standards, for example, in 3G with the S-UMTS standard [4]. There have also been attempts to standardize hybrid terrestrial and satellite connectivity through the Auxiliary Terrestrial Component Specifications in the United States, Canada, Europe, and Asia and, in China, the Satellite and Terrestrial Multi Service Infrastructure [5].

At a 3GPP Technical Standards Group (TSG) meeting in March 2017, it was agreed that a 5G and nonterrestrial networks (NTN) study would be produced within the 3GPP Release 15 standards process (New Radio NTN, NR.NTN). The sponsors included Motorola, Sepura (emergency service radio), the Indian Institute of Technology, Avanti, Mitsubishi, China Mobile, and Airbus Group.

The standards work extends across six domains:

- The support of 5G connectivity via satellite within 3GPP TR23.799;
- The higher availability requirement within 3GPP TR22.862;
- The wide area connectivity requirement within 3GPP TR22.863;
- The satellite access requirements within 3GPP TR 22.864;
- The 5G connectivity using satellites use case of 3GPP TR 22.891;
- The satellite extension to terrestrial within 3GPP TR 38.913.

However, our specific interest in this chapter is the modeling activity associated with these work streams.

There are five proposed deployment scenarios including geostationary, nongeostationary, and subspace (high-altitude platform systems) with a range of considered FDD bands including 2 GHz, 6 GHz, 20 GHz, and 30 GHz deployed either as bent pipe or with onboard processing with channel bandwidths of 20 MHz, 80 MHz, and 800 MHz, outdoor or outdoor/indoor (subspace) with either fixed or moving beams.

Note there is a double Doppler effect that has to be accommodated. LEO satellites are traveling at a speed of about 28,000 km/hr (7.7 km/sec) depending on their orbit altitude. The satellite Doppler is a known constant; the moving object with which the satellite is communicating will typically be moving at different and variable velocity. Although Doppler might be considered problematic, it is a well understood effect exhibiting itself as an increase or decrease in frequency depending on whether the objects are traveling towards each other (an increase in frequency) or away from each other(a decrease in frequency). The strong Doppler signature of LEO satellites can be used to provide precise positioning and location services, so it can be regarded as an asset rather than a problem.

Within the group, Hughes Network Systems are providing inputs to the free space loss assumptions. The assumptions highlight the need for additional gain from the antennas at both ends of the link, with a particular need to address antenna design issues for user and IoT terminals in Ka-band.

For example, if the L-band frequency loss is assumed as 4 dB, then the relative loss for S-band will be 6 dB and 29 dB for 28 GHz.

For a GSO with a path link distance of between 35,788 km and 41,679 km, the distance loss will be 91.1–92.4 dB. The total loss will be >187.5 dB at 1.6 GHz and <212 dB at 28 GHz.

For a LEO at an orbit altitude of 600 to 1,500 km, the distance loss will be 55.6 dB to 63.5 dB. The total loss at 1.6 GHz will be >152 dB and <185 dB at 28 GHz.

A high-altitude platform at a height of between 20 km and 40 km will have a distance loss of 26–29 dB with a total loss in the range of <122.4 dB at 1.6 GHz and <150.5 dB at 28 GHz.

Gain is a function of beamwidth, but the link budget will also be affected by other factors including ground reflection. The path loss is significantly greater in a GEO network and lowest for a high-altitude platform systems network due to the shorter path length. The path loss from a terrestrial base station to a user a few meters away will be theoretically better than any nonterrestrial connection but by less of a margin than you might expect particularly at millimeter frequencies where nonline-of-sight losses and surface absorption absorb significant amounts of RF signal energy.

Size is the great savior. Given that antenna element spacing is inversely proportional to the carrier frequency, a 30-GHz antenna will be 10 times smaller than a 3-GHz antenna. A 28-GHz antenna will be 10 times smaller than a 2.8-GHz antenna.

Put another way, if the antenna is kept at the same size and the element count is increased, then the beamwidth will reduce proportionately, increasing isotropic gain and reducing visibility to unwanted signal energy. A 21.9° half-power beamwidth antenna will have a gain of 18 dB. A 1.23° fractional beamwidth antenna will have a gain of 43 dB.

Note that narrower beamwidth antennas pointed directly upwards at the sky will also see a higher ratio of required power to scattered power and will be less affected by ground reflections.

Various study groups are presently working on shorter wavelength terrestrial and satellite channel characteristics including ITU-R P.681 and 682 and 1853 managed by the European Space Agency [6].

6.4 Back to Earth: 5G Antenna Trends

In my last book, *5G Spectrum and Standards*, I covered 5G antennas featuring products from Blu Wireless at 60 GHz, Huber and Suhner (millimeter-band antennas), antenna tilt techniques in the sub-1-GHz band from Quintel, and, for good measure, the Ryle radio telescopes at the Mullard Radio Astronomy Observatory and automotive radar antennas.

In this next section, we review the technology innovations and new products that have emerged in the 2 years since the last book was written and published.

We talked briefly about antennas for backhaul, but increasing network density and the growing recognition that 5G backhaul operational and capital costs need to be constrained have placed increased attention on backhaul connectivity so that seems like a good place to start.

6.4.1 5G Backhaul

The band-naming regimes are very confusing. We have probably just about got our heads around the IEEE 521-1984 radar band designations with Ku-band at 12–18 GHz, K-band at 18–27 GHz, Ka-band at 27–40 GHz, V-band at 40–75 GHz, and W-band at 75–110 GHz.

In fixed point-to-point hardware specification sheets, you will also come across bands described using the WR22 waveband designation [7], for example, the bands at 40 GHz, which are described as Q-band, and the WR12 waveguide designation, for example, the 71–76-GHz and 81–86-GHz allocations known as E-band [8]. This is because these products have typically been implemented as wave guides and horn antennas manufactured to very close tolerances.

A typical product is shown in Figure 6.4. This is a dual-polarized horn antenna covering 50–75 GHz with 15-dBi nominal gain and a half-power beamwidth of 28° in the E plane and 33° in the H plane [9].

This is variously described as a V-band antenna or WR-15 waveguide. For waveguide-naming conventions, see [10].

There are a bewildering number of fixed point-to-point products available across a bewildering range of frequencies and channel bandwidths. Essentially, these are hand-crafted products built in hundreds or thousands rather than millions or billions. A nicely documented summary of a contemporary fixed point-to-point product range has been produced by RF.com (Figure 6.5) [11].

Table 6.1 provides a comparison of the relative gain available at Q-band and E-band for a 2-foot dish antenna at Q-band and E-band across channel bandwidths from 250 MHz to 2 GHz and related maximum throughputs per channel. As can be seen, significant additional gain can be achieved at E-band due to the additional aperture gain available from the dish antenna at these shorter wavelengths. There are additional propagation losses at E-band and receiver sensitivity may be a bit less than a Q-band receiver due to a higher noise floor, but it is possible to get significantly higher throughput over an E-band link without a significant loss of range with 10 Gbps being the highest claimed throughput on this particular RF hardware platform. Throughput can be increased by using higher-order modulation, but range would decrease. As a rule of thumb, every doubling of modulation state will take 3 dB off the link budget.

Figure 6.4 Dual-polarized horn antenna. (With thanks to Sage Millimetre.)

Figure 6.5 RF Com Dish with integral transceiver for point to point backhaul in a 4G or 5G network. (With thanks to RF.com.)

Table 6.1
Gain and Range from RF.com Q-Band and E-Band Dish Antennas

Band	Q-Band				E-Band		
Frequency	40.5–43.5 GHz				71–76/81–86 GHz		
Throughput	Up to 10 Gbps full-duplex						
Channel bandwidths	250/500/750/1,000/1,250/1,500/2,000 MHz						
Modulation	QPSK to 256 QAM						
Max distance 2-ft antennas, clear sky	Up to 20 km (12 mi)						
Antennas: gain and beam width	Cassegrain with radome						
	44 dB, 0.7°, Q-band 40 GHz				51 dB, 0.35°, E-band (70/80 GHz)		
QPSK link budget by channel bandwidth	183 dB at 250 MHz, 180 dB at 500 MHz, 178 dB at 750 MHz, 177 dB at 1,250 MHz, 176 dB at 1,000 MHz				197 dB at 250 MHz, 194 dB at 500 MHz, 192 dB at 750 MHz, 191 dB at 1,000 MHz, 190 dB at 1,250 MHz, 189 dB at 2,000 MHz, 188 dB at 1,500 MHz		
Max. throughput Q and E bands	250 MHz	500 MHz	750 MHz	1,000 MHz	1,250 MHz	1,500 MHz	2,000 MHz
Mbps	1,750	3,450	5,290	7,045	7,430	8,940	10 Gbps

Source: RF.com.

6.4.2 Self-Backhauling/In-Band Backhauling in 5G

It has been recognized that 5G will be deployed into urban environments at a density which would mean that separate RF backhaul or fiber backhaul will be uneconomical. The economic cost is a consequence of the sunk cost of fiber and or hardware cost of separate point-to-point dishes and transceivers.

The performance cost is a consequence of any demodulation/modulation or channel coding added at the transition points between the 5G physical layer and fiber backhaul. RF over fiber [12] is a partial answer to this, but an increasingly promoted option is to implement self-backhauling in which the same radio resources are available for users and backhaul. This is sometimes known as in-band backhauling [13].

The backhaul market is a market that new high-count LEO satellite operators such as OneWeb are keen to penetrate. The advantage of self-backhauling is that a terrestrial operator can reuse RF hardware base station resources across the user plane, control plane, and backhaul plane, but this implies there is a requirement to go around corners. This may or not be convenient depending on where base stations can be sited. Satellite operators presently have a small percentage of the mobile broadband terrestrial backhaul market, less than 1%, with much of that in hard-to-reach deep rural areas.

Increasing satellite connectivity into local ultradense urban backhaul competing against self-backhauling would be dependent on meeting latency constraints at a cost equal to and preferably lower than in-band backhauling, bearing in mind that the in-band option amortizes hardware and bandwidth costs across users and base station-to-base station backhaul. The one advantage that satellites have, particularly nearly always nearly overhead satellites, is that there may be a higher probability of a clear direct line of sight to all the base stations within a confined area. This would avoid the mesh protocol overheads incurred in self-backhauling. Note that latency introduced by mesh protocols will be variable with the variability dependent on the local base station deployment topology. Satellites may introduce additional latency (see Chapter 2), but the latency, at least from nearly always nearly overhead LEO constellations, will be essentially constant, which would mean that any higher-layer protocol overheads into, across, and out of the backhaul plane could be minimized.

6.5 Innovation in Terrestrial 5G and Nonterrestrial Network Antennas

6.5.1 Steerable Mechanical Antennas

Dishes are an efficient option for achieving directional gain in terrestrial backhaul networks and in satellite networks. They can be mechanically repointed to send and receive signals in other directions, although this is a relatively slow and cumbersome process. Mechanical beam steering has been used in radar systems since World War II. If the mechanical pointing failed, the truck could be driven around in a circle, which would change the azimuth, although not the elevation.

6.5.2 Electrically Steerable Antennas Using Conventional Components and Materials

In the 1990s, companies such as Arraycomm [14] and later Quintel [15] began to introduce electronically steerable antennas in which the phase offsets between antenna elements are changed to create nulls to mitigate interference and gain to improve directional range and throughput. At lower frequencies, particularly bands below 2 GHz, these antenna arrays can be large and the weight and wind loading can add significantly to mast costs, but they do solve specific interference problems in specific places.

Electrically steerable antennas at higher frequencies and shorter wavelengths have the advantage that elements can be closer together and it becomes more viable to build flat panel antennas that are compact enough to survive high winds and the occasional or not so occasional hurricane. They can also switch beam pattern far faster (milliseconds or microseconds and potentially picoseconds) than mechanically pointed arrays (seconds). Flat panel electrically steered antennas have now become widely deployed in military radio and radar systems and automotive radar (see Chapter 10, pages 264–273, in *5G Spectrum and Standards*). They can be built using conventional components and materials with elements of various lengths so that the antennas are steerable and wideband.

These antennas can also be constructed using a class of materials called metamaterials.

6.5.3 Electrically Steerable Antennas Using Metamaterials

Metamaterials (meta, from the Greek, meaning beyond) are materials that have properties that are not found in nature and are usually arranged in repeated patterns at scales that are shorter than the wavelengths of the medium with which they are intended to interact (Figure 6.6). It is therefore the structure and its shape and orientation and arrangement as much as the base material that influences the performance and behavior of the device.

It could be argued that PIFA antennas [16] are a precursor to metamaterials and are one example of size-efficient shapes and structures in conventional antennas coupled with innovative ground planes. However, metamaterials are more complex and elaborate. As with electrically steerable conventional antennas, metamaterial-based antennas are becoming widely used in military radio and radar systems, including wideband radio systems scaling from UHF to K-band. They can enhance and block, absorb, and bend electromagnetic waves. As with conventionally structured antennas, they cannot do all these things at once so interpret marketing material and specification sheets with a measure of care.

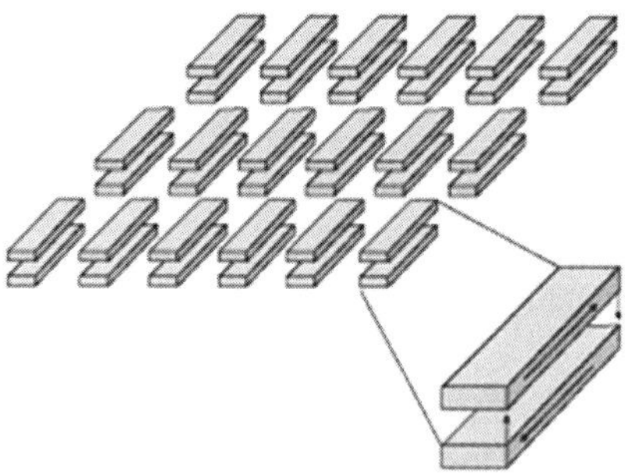

Figure 6.6 Metamaterials. (Image courtesy of Kymeta.)

6.5.4 Metamaterial Antennas Combined with Electromagnetic Bandgap Material

A second class of material known as electromagnetic bandgap (EBG) material [17] can be combined with metamaterials to mitigate the distance separation issue of antennas at lower frequencies.

Developed by the U.S. Army Research Laboratory at the University of Michigan, these materials are claimed to realize an antenna in S-band at 2.72 GHz with a 3-cm physical separation but with a 42-dB isolation between the antennas, 24 dB above the isolation achievable with conventional antenna materials. Put another way, realizing a 3-cm separation distance using EBG material is acclaimed to be equivalent to a meter separation using conventional materials.

Xerox PARC has used these materials and manufacturing techniques to develop an RF beam steering platform. The platform is being engineered by Metawave into a range of antenna products and applications. Figure 6.7 shows the structure of the device realized as a 32-element array. The antennas are being developed for the automotive industry (Figure 6.8) [18].

The same hardware and software can be repurposed for 5G as shown in Figure 6.9. Note that this is effectively a progressive point-to-point network, although with the user and supported device unaware of the process of beam-to-beam handover.

Kymeta has a similar product but more generally applied for connectivity in Ku-band, emphasizing the inteference null form capabilities of the device (Figure 6.10). Kymeta has a number of case studies of connected transport appications using their antennas including projects in the United States with Toyota [19].

Figure 6.7 A 32-element steerable antenna from Metawave.

Figure 6.8 Metawave scanning radar for autonomous driving.

Figure 6.9 Metawave antennas for 5G progressive point to point.

6.5.5 Active Conformal and Flat and Almost Flat Antennas

Conformal antennas are antennas that can be molded to any shape, for example, a car, truck, military tank roof, superyacht bridge, train roof, aircraft hull. Often, they are almost flat with a small amount of curvature. A conformal antenna for a completely flat surface, unsurprisingly, will be completely flat.

Phasor Solutions [20] presently produce these antennas for high-end luxury yachts or highly specified military use (Figure 6.11), but it can be imagined that these would be effective as a flat roof-mounted antenna on a car roof

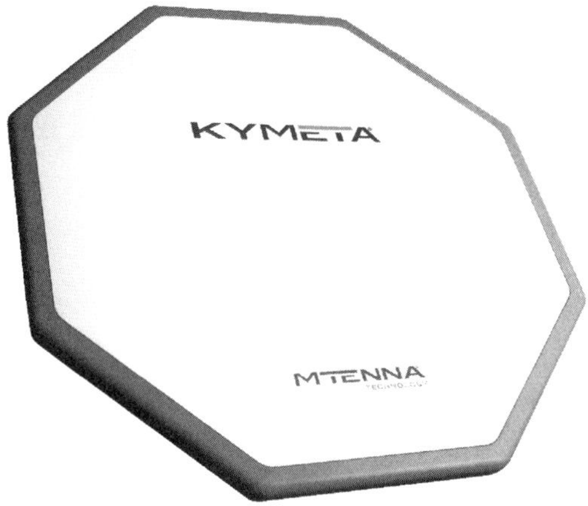

Figure 6.10 Kymeta Ku-band metamaterial antenna.

pointing at satellites. The active beam steering allows unwanted angular power to be nulled out but also enables RF power to be delivered and received across a wide range of elevation angles. For instance, a vehicle at high northern or southern latitude served by a geostationary satellite would be focused on an elevation close to the horizon and be configured to have minimal visibility to unwanted signal energy coming, for example, from LEO satellites directly overhead.

A six-module phasor antenna specified to receive (10.7–12.75 GHz) and transmit (14.0–14.5 GHz) in Ku-band has an instantaneous bandwidth of 125 MHz and will handle up to 500W across a temperature range of −55°C to +85°C and weighs 12 kg.

Table 6.2 compares phasor antenna performance against equivalent dish antennas.

Figure 6.12 shows an example of a curved conformal antenna.

6.5.6 Active and Passive Conformal Antennas

Active antennas arrays are presently inherently expensive as each element has its own RF power and low noise amplifier and associated filter and matching networks. They could also be required to work across a wide temperature range, for example, +125°C in automotive roof-mounted applications (compared to the +85 specified in the example above). This is a hard-to-realize cost economically and can also result in performance degradation caused by frequency drift with temperature and noise rise from the heat energy absorbed by the device.

An alternative is to construct conformal antennas with elements that are mechanically and electrically arranged to look directly upwards and nowhere else. Effectively, this is a passive antenna with multiple antennas with phase

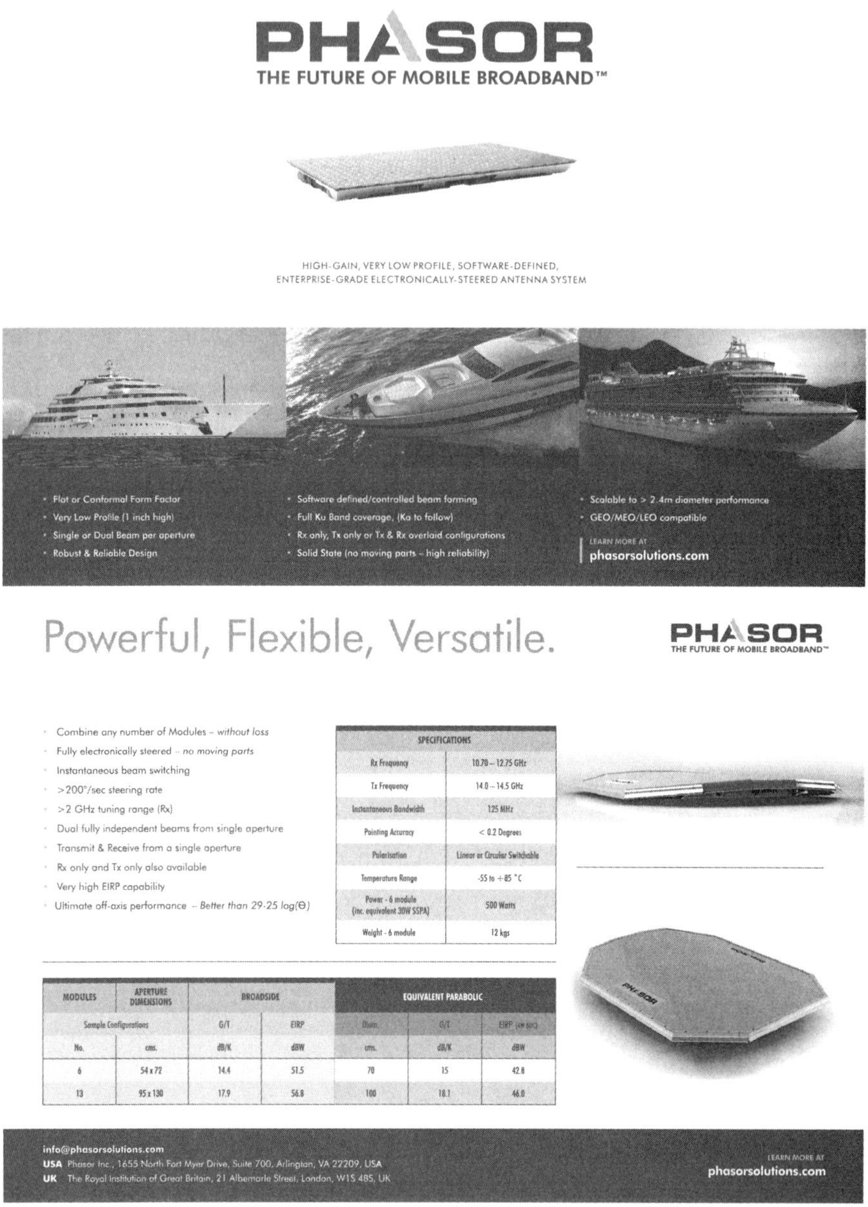

Figure 6.11 Phasor Solutions' active conformal antenna.

offsets created using passive delay lines and one RF power and low noise amplifier, which can be remotely mounted. These devices do not have adaptive capabilities and look at the same bit of sky, but they are lower-cost, thinner, and less temperature-sensitive. They could be very adequate when used with high-count LEO always directly overhead constellations.

Table 6.2
Phasor Antenna Performance Compared to Dish Antennas

Modules Number	Aperture Dimensions cm	EIRP dBW	Equvalent Dish Diameter cm	EIRP dBW
6	54 × 72	53.6	70	42.8
12	72 × 108	59.6	100	46
27	126 × 144	66.6	150	49.6

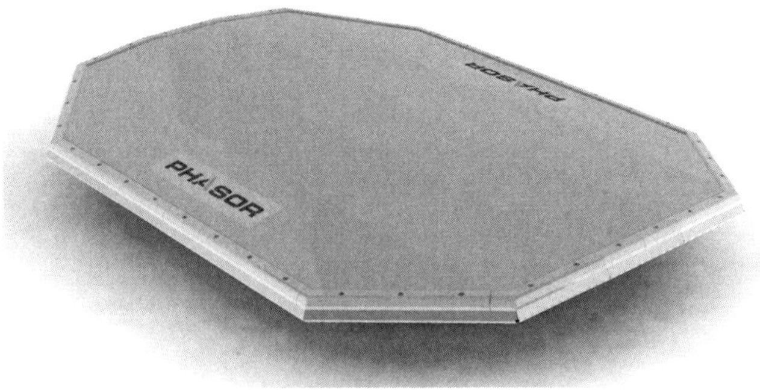

Figure 6.12 Phasor conformal (curved) antenna.

6.5.7 Active Electronically Steered Array Antennas for Military Radar, SATCOM, and 5G Terrestrial and 5G Backhaul Applications

The principle of an active conformal flat panel antenna array is that it can detect and analyze the angular power received into the antenna both in terms of elevation and azimuth and therefore determine where RF energy should be focused on the return path.

This is similar in principle to radars used since World War II, although in these early legacy radar systems, the return path is anti-aircraft fire. Modern anti-missile missile systems provide contemporary leading-edge examples of how digital processing can work out the angle of arrival and the trajectory and speed of a close or distant object. Switching speed can be in the order of nanoseconds (a nanosecond is one thousand-millionth of a second). These antennas systems are referred to as active electronically steered array (AESA) radar. When used in communication systems, they are known simply as AESA systems.

AESA radar and communication systems are manufactured by a wide cross-section of the military systems supply chain including Raytheon, Boeing, Lockheed Martin, Northrop Grumman, and BAE Systems. IBM, Intel, and Si-Beam are also invested in the sector.

Automotive radar (starting on page 264 in *5G Standards and Spectrum*) essentially does the same calculation though with a different desired outcome (to miss rather than hit the object ahead).

Figure 6.13 shows an active antenna integrated circuit (IC) product from Anokiwave that can be used across SATCOM, radar, and 5G terrestrial applications. See Figure 6.14 for the Anokiwave product range.

The ability to amortize development costs across these multiple markets, both in terms of spatial processing algorithmic development and optimized hardware architectures, is a significant advantage. The 5G market offers substantial volume opportunities but only at a price point several orders of magnitude below more specialist applications. A $3,000 antenna on a superyacht or a $30,000 antenna on a fighter jet or tank cannot be translated directly to a base station under $10,000 or a Wi-Fi access point under $100. There are also different design requirements. Switch speed in radar systems, for example, is a critical performance parameter.

6.6 4G and 5G Terrestrial AESA Systems: Flexible MIMO

This brings us to a discussion of how terrestrial MIMO systems might evolve over the next few years. Each of the major Tier 1 vendors (Huawei, Ericsson, and Nokia) has invested substantial effort in developing MIMO systems for

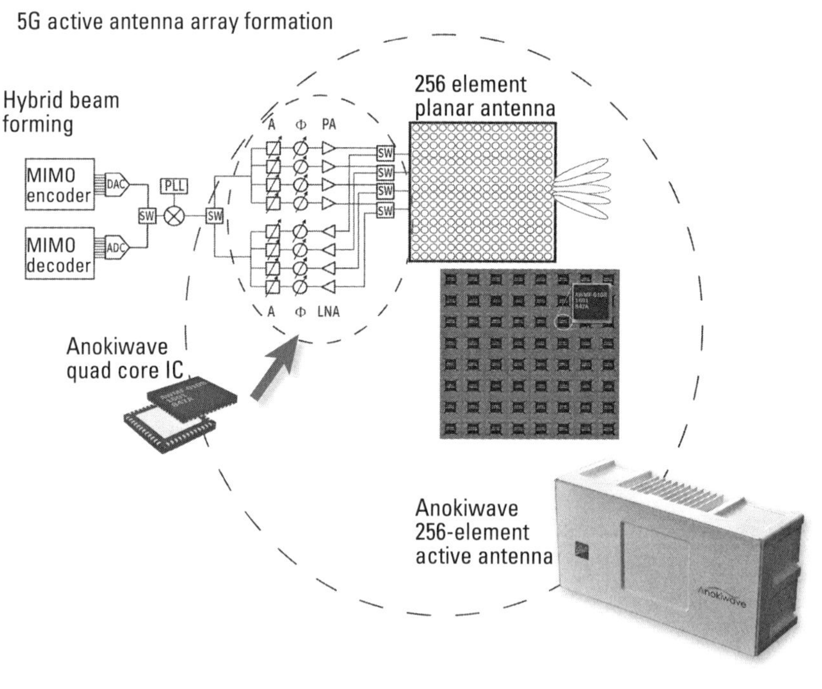

Figure 6.13 Anokiwave active antenna integrated circuit for active electronically steerable array antennas [21].

Market	Product Family	Part Number	Description
5G Communications Antenna Arrays	Ka-Band Silicon IC Solutions	AWMF-0135 AWMF-0108 AWMF-0123 AWMF-0125	26 GHz Quad Core IC 28 GHz Quad Core IC 39 GHz Quad Core IC 39 GHz Quad Core IC
	Ka-Band Active Antennas	AWMF-0129 AWA-0134	28 GHz 64 Element Innovator Kit 28 GHz 256 Element Innovator Kit
mmW Communications and Sensing Applications	Ku- and Ka-Band Si Core ICs	AWMF-0117 AWMF-0116	Ku-Band Multi-Function Core IC Ka-Band Multi-Function Core IC
RADAR and Communications Active Antennas	X-Band Silicon Core IC Solutions	AWS-0101 AWS-0103 AWS-0104 AWS-0105	Dual Beam Low NF Quad Core IC Dual Beam High IIP3 Quad Core IC Single Beam Low NF Quad Core IC Single Beam High IIP3 Quad Core IC
RADAR and Communications Active Antennas	X-Band Front End ASIC Solutions	AWMF-0106	Medium Power Front End ASIC
SATCOM Active Antennas	K and Ka-Band Silicon Core IC Solutions	AWS-0102 AWMF-0109 AWMF-0112 AWMF-0113	4-element Rx Quad Core IC (K-Band) 4-element Tx Quad Core IC (Ka-Band) 8-element Rx Octal Core IC (K-Band) 8-element Tx Octal Core IC (Ka-Band)
SATCOM, AESA, Point-to-Point Communications	Ka-Band III/V Front End Solutions	AWP1102	3W High Power Amplifier
Point-to-Point Radio Communications	E-Band III/V Front End Solutions	AWP-7176 AWP-8186 AWL-7186	High Power Amplifier MMIC/71-86 GHz High Power Amplifier MMIC/81-86 GHz Low Noise Amplifier MMIC/71-86 GHz

Figure 6.14 Anokiwave product range.

high throughput 4G and 5G networks. The challenge is to make these platforms flexible and fast enough to respond to changing channel conditions including changes in the angular direction of arrival of wanted and unwanted signal energy but delivered at price points several orders of magnitude lower than the military radar systems described above. The products are described as flexible MIMO.

6.6.1 Automotive AESA

Automotive radar has a similar pricing and cost issue, although the automotive supply chain seems to be efficient at bringing innovative, low-cost, high-performance radar products to market.

This is partly due to the scale value of the automotive industry. We revisit this in later chapters but consider:

- The number of planes in the world: 11,000;
- The number of trains in the world: >20,000;
- The number of ships in the world: 50,000;
- The number of cars in the world: 1.5 billion;
- The number of people in the world: 7.5 billion;
- The number of connected devices in the world: 10 billion.

Planes, trains, and ships have a higher value than cars, but the car market is bigger in volume terms. A high-end car valued at >$50,000 will typically have 11 forward rearward and sideways facing radars. The economic value of people and devices is a philosophical debate which we do not have time to address in this chapter, but antenna vendors need to find a sweet spot between volume and value in their development and market plans. The value of radar in a car is potentially the value realized from making a car impossible to crash, but to be fully effective, all cars need to be impossible to crash which may take a while.

6.6.2 Some Nokia Examples of 5G Flexible MIMO Antenna Arrays

5G vendors may have an ambition to service all the above markets but are generally dominant in people and device connectivity. As discussed in various other parts of this book, people and devices have different connectivity requirements (and different amounts of money to spend), which means that antennas must do many different things and can be hard to optimize for general tasks. The starting point is to decide whether functions such as beam forming are implemented in the analog or digital domain or a combination of both.

Note the trade-offs between coverage and capacity and power consumption and bandwidth limitations. The all-digital option, for example, is the most flexible but constrained in terms of channel bandwidth due to the limitations of present digital signal processing technology.

6.7 Beam Frequency Separation

These antennas will also change frequency planning in 5G and satellite networks. In first, second, and third generation cellular networks, base station-to-base station interference was managed in the frequency domain with adjacent base stations using different channels within each passband. In 4G and 5G networks, there has been a progressive move to single-frequency networks where the same channels are available from all sites with interference managed in the time domain (intercell interference coordination).

In 5G, users are supported by their own dedicated beam and beam-to-beam separation is achieved spatially.

In satellite networks, a beam usually services a geographical area with multiple users, analogous to a traditional cell pattern in a terrestrial cellular network. Users download all traffic within, for example, a 250-MHz channel within a 3.5-GHz passband and then just extract the packets with the correct IP header address. This is adequate but not particularly efficient (additional clock cycles are consumed due to processing the packets that are then discarded).

As constellation density increases, particularly in high-count LEO constellations (including CubeSats), and as the number of beams and the directivity of those beams increases on larger satellites, it becomes increasingly feasible to support individual users within those geographic areas.

6.8 Plasma Antennas

We have described antennas that use copper and similar high-conductivity materials to construct structures that can translate RF signal energy into an induced voltage.

There is another alternative known as a plasma antenna. Plasma antennas, as the term implies, are RF structures that use plasma as the guiding medium for achieving resonance with modulated radio carriers. First patented in 1919, they may be having their millennial moment in mobile communications.

There are advantages and disadvantages associated with plasma antennas that can be summarized as:

- The length of an ionized filament can be changed rapidly, thereby retuning the antenna to a new frequency.
- The antenna can be turned off to make it electrically invisible for the purpose of reducing its scattering signature and eliminating its coupling and interference with other nearby antennas.
- The use of plasma adds complexity to the antenna design.
- Equipment for establishing and maintaining the ionization must be provided.
- The glow from the plasma increases its visible signature, and plasma decay generates noise.
- A plasma antenna can be established in air at atmospheric pressure by using lasers, high-power microwave beams, or ultraviolet rays. A plasma can also be generated within a tube containing a noble gas (a gas that is

unreactive except under extreme conditions), for example, neon and argon. Methods that use a tube require less energy to excite and maintain the plasma state, because the gas is pure and the presence of the tube prevents dissipation, but the use of a tube increases the antenna weight and volume and makes the antenna less durable.

There are demonstration products available in bands targeted for 5G, including products from Plasma Antennas Limited. These devices can be stacked to form and steer beams in azimuth and elevation to form multiple beams.

6.9 Flat VSATs and Their Role in LEO, MEO, and GEO Interference Mitigation

The narrative of this chapter so far has been that antennas can facilitate in-band cosharing between the satellite industry and 5G and facilitate band-sharing and mixed constellation delivery from LEO, MEO, and GSO satellites.

To date, geostationary operators and MEO operators have been challenging the interference models put forward by the NEWLEO entities.

Frequency scaling can be used to increase the number of antenna elements with an extra 6 dB of gain available for every doubling of antenna count. Alternatively, a lower antenna count could be used to realize a small antenna with the size reducing with frequency, so, for example, the Apple Series 3 watch with an LTE transceiver has an antenna integrated into the screen. If this was scaled to a full-size smartphone screen, the result could be a 32-element antenna.

Now imagine that we take this device and place it on a table so that it looks at the sky. The antenna array can be passive or active. In a passive array, the phase offsets between the elements can be set to provide a fixed nonvariable cone of visibility that looks directly upwards. The narrower the cone of visibility, the higher the uplink and downlink gain. A narrow cone of visibility will also that mean unwanted signal energy from satellites not directly overhead will be nulled out. For larger form factor antennas, for example, a large passive flat panel array placed on a flat roof, the gain will be higher, and the rejection of unwanted signal energy will be higher.

On or near the equator, the passive phase array flat antenna looking directly upwards will have visibility to geostationary satellites and any MEO and LEO satellites that happen to be passing overhead and a selection must be made as to the strongest available signal and/or signal with the lowest latency, which will generally be the LEO.

Moving to higher latitudes north or south will mean that visibility to LEO and MEO satellites becomes progressively more dominant.

Passive flat panel and passive conformal antennas (antennas shaped to the host structure) are therefore a potentially low-cost mechanism for separating out wanted from unwanted signal energy at higher latitudes by providing a cone of visibility on the transmit and receive path to satellites that are directly overhead. By definition, these will be the satellites with the shortest path through the atmosphere and therefore the lowest latency and lowest path loss.

Active electronically steerable flat panel array antennas perform a similar function but in a different way. An active electronically steerable flat panel array can scan from horizon to horizon and select the best path available from any visible satellite, which could be a GSO, MEO, or LEO. For example, in a thunderstorm, a directly upwards path might suffer high rain fade and a better link might be available from a GSO at a lower elevation. Choices can also be made based on required latency or throughput or cost. Having acquired a beam from the best available satellite, the AESA antenna can null out the signal energy from all other sources. These active antennas can be available as flat panel or conformal. However, they are more expensive because each antenna element must have its own RF amplifier, low noise amplifier, and switch path. By comparison, a passive antenna has one RF power amplifier, one LNA, and one switch path, and the RF power amplifier and LNA can be mounted remotely. In an active antenna, the component cost scales as a function of the number of antenna elements but also the components are temperature sensitive and will normally be specified from −55°C to +85°C. This is not sufficient for applications such as the automotive industry where +125°C will generally need to be accommodated.

These active and passive flat antennas can be considered as a kind of flat very small aperture terminal (VSAT). A VSAT dish will be pointed at a particular bit of sky and will provide sufficient uplink and downlink gain to deliver high data rates, for example, for business to business corporate data networks. All we have done is to recreate the function of a VSAT dish but with the capability to point either directly upwards (passive flat/passive conformal) or anywhere in the sky depending on satellite availability (active flat and active conformal).

The antennas can either be high element count (256, 512, 1,024 elements) flat panel or conformal arrays for mounting on cars and trucks and any large object that is large and either moving fast or slowly. We call these applications the Internet of Fast-Moving Objects (IoFMO) like trains, planes, and fast cars, and the Internet of Slow-Moving Objects (IoSMO) like milk floats, tanks, and cars in a slowly moving traffic jam, but they are often described in the literature and standards documents as Earth stations in motion (ESIM). Note that fast-moving objects consume disproportionate amounts of signaling bandwidth in terrestrial networks and are often more efficiently served from directly overhead satellites.

The other application is the Internet of Stationary Objects (IoSO), also known as stationary Earth stations: Wi-Fi-enabled trash bins in Singapore are one example. Counterintuitively, stationary objects can be hard to service from a terrestrial network, particularly when the object is nonline of sight in higher-frequency bands. At least a moving object will have visibility to a 4G or 5G base station some of the time.

6.10 Scaling Flat VSATs by Wavelength and Size

Table 6.3 shows how flat and conformal VSATs scale by wavelength, size, and number of antenna elements and how this could map to a range of potential applications.

As referenced earlier, the size of the antenna can be kept constant with the number of elements increasing as frequency increases/wavelength reduces. This will have the effect of providing more interference rejection and the ability to provide a narrower cone of visibility directly upwards.

Alternatively, the size of the antenna at any center frequency can be increased to support additional antenna elements. In either case, there is an additional 6 dB of gain for every doubling of the number of elements, which can be realized as throughput gain, range gain, and interference rejection or some combination of these.

In terms of application, 4, 6, or 8-element arrays would generally be targeted towards small wearables, and small form factor IoT, 32, 64, and 128-element arrays are probably the sweet spot for smartphones and the higher element count arrays are probably best suited for use as conformal antennas for cars, trucks, boats, and planes.

6.11 Can Flat VSATs Be Produced at Low Cost?

Active conformal antennas available today and referenced earlier in this chapter are relatively expensive and companies such as Phasor focus on high value applications such as super yachts and military applications and planes and trains.

However active and passive flat panel arrays could potentially be manufactured either using LCD display production lines or solar panel production lines and the antenna elements could be embedded in an LCD screen (to provide terrestrial connectivity in the horizontal plane for example to devices in a living room). If embedded into a solar panel the devices would be cross-amortized across two applications domains (photon capture and electron capture) and could be pointing upwards to provide outdoor connectivity to LEO, MEO, or GSO satellites. There will also be localized power available.

6.12 The 28-GHz VSAT Smartphone

For smartphones, a 32-element array implemented, for example, at 28 GHz, could be embedded in a smartphone display. When used as a handheld device, the smartphone will have visibility to terrestrial 4G LTE or 5G networks. Outdoors and in remote areas, a user would simply put the smartphone on a flat surface pointing at the sky to provide visibility to LEO, MEO, and GSO constellations.

6.13 Multiband Flat and Conformal VSATs

As stated above, metamaterial-based active and passive flat and conformal antennas can be wideband with elements that can be linked together to provide wavelength resonance to lower frequencies, for example, down to VHF. Note that we have stated that within the present terms of reference for bands for study for satellites, the bands of interest scale from 138 MHz (Orbcomm OG2) through high throughput satellites in the K-bands to V and W-band (E-band) for superhigh throughput satellites.

The element count will reduce as additional line lengths are switched in so a 1,024-element array at E-band (the 72–77 GHz, 81–86 GHz duplex) would be a 4, 8, or 16-element array at UHF. Note that a passive wideband antenna will have to have an active switch matrix.

The 5G bands of interest scale in a similar way, starting at 450 MHz (Band 31) and ending at the top of E-band (92–95 GHz).

6.14 What Physical Layer Should Satellites Use?

In Chapter 10, we review satellite and 5G standards. The important difference between 4G/5G and satellite is that the satellite downlink is optimized for power efficiency, and therefore most satellite constellations use phase amplitude shift keying (PASK) rather than the QPSK used in 4G (and probably 5G). The multiplexed 4G/5G composite waveform, for example, has considerable amounts of AM. It is reasonably spectrally efficient but not power-efficient. However, an LTE or 5G front end in a smartphone, smart watch, or IoT device would have no problem accommodating an APSK-modulated composite multiplexed waveform providing the channel bandwidths and passbands are the same for both systems, so, for example, a 250-MHz channel raster in Ku-band at 12 GHz and a 250-MHz or 500-MHz channel raster at Ka-band and a 500-MHz channel raster at V-band (37.5 to 40 GHz, 51.4 to 52.4 MHz) and a 1-GHz channel bandwidth in the 5-GHz passband in E-band FDD (71–76, 81–86 GHz) and TDD E-band (92–95 GHz) would be completely compatible.

Table 6.3
Scaling Flat and Conformal VSATs by Wavelength and Size

Flat and Conformal VSATs		
Scaling by Wavelength		
Meter band 300 MHz to 3 GHz	Centimeter band 3–30 GHz	Millimeter band 30–95 GHz
Number of phase array elements		
4 8 16	32 64 128	256 512 1,024
Scaling by Size		
Small	Medium	Large
4 8 16	32 64 128	256 512 1,024
Throughput gain Range gain Interference rejection Example applications		
Small wearables	Smartphones	Cars, trucks, boats, and planes

6.15 Band-Sharing 5G with High Throughput Gigabit Satellites at 12 GHz and 28 GHz, with Very High Throughput Terabit Satellites at 40/50 GHz and Superhigh Throughput Petabit Satellites in E-Band

Band sharing is the natural consequence of having flat VSATs that can discriminate between horizontal 5G terrestrial network availability and directly overhead LEO, MEO, and GSO vertical coverage. In-band 5G backhaul can be included as well. This is the Power of Five delivery model in which five delivery systems, 5G terrestrial, 5G in-band backhaul, LEO, MEO, and GSO all coshare the same passbands.

This avoids 10 years of ferocious argument between the satellite industry and 5G industry. It also allows existing incumbents in all the other bands to stay where they are, avoiding 10 years of arguments with other stakeholders including military radio and radar, deep space communication, and radio astronomy. It delivers scale benefits to the satellite industry including the opportunity to capture some of the consumer market connectivity value of the smartphone and emerging wearables market. It solves many of the network density and deep rural cost issues of the 5G community and the issue of "not spot" 4G and 5G coverage in urban and rural environments.

Note that it also allows the satellite industry to continue to serve their traditional high added-value users with high-end, high-performance, high antenna count active conformal antennas. Table 6.4 summarizes this wireless nirvana. Note the differentiation between servicing the IoFMO, IoSMO, and IoSO.

6.16 Flat VSATs and Wireless Wearables?

It is all very well putting a 16 or 32-element array antenna into a smartphone display, but some people still make phone calls on their phone, which would mean that the 5G antenna array would be pointing directly towards the user's head.

There are several answers to this, including putting the antenna array somewhere else on the phone, but it is also possible that form factors other than the smartphone will emerge. The Apple 3 Wireless watch may be the start of an LTE-based wearables mass market that would open up new form factor opportunities. Remember from earlier in the chapter that flat VSATs are based on a combination of metamaterials (materials with conducting properties constructed in shapes that provide wavelength resonance performance not achievable from conventional antenna structures) and electronic bandgap materials that mitigate unwanted coupling between closely spaced antenna elements.

It is plausible that these material and manufacturing techniques could be incorporated into new added value clothing ranges, the VHF to V-band action vest with an antenna array printed on the chest of the vest; a Superman logo footprint would do nicely.

The disadvantage with this is that Superman could only talk to a satellite directly overhead by lying down and looking at the sky. An alternative would be the Batman broadband connectivity suit with additional head-mounted, dual-polarized antennas, although the specific absorption rate (SAR) limits would need to be calculated carefully [23].

What other problems can flat VSATs solve?

6.17 The Role of Flat VSATs: Solving the Ground Gateway Interference Problem and Cost Problem

The other tension point between GSO, MEO, and LEO operators is the Earth gateway interference issue. Feeder links at 18 GHz, for example, are common to GSO, MEO, and LEO constellations and any of these can pour unwanted signal energy into Earth gateways supporting other systems.

Gateways are expensive in terms of the real estate they occupy, their power requirements, and the hardware (mechanically steerable dish antennas) that must be deployed. The number of gateways can be reduced by using intersatellite switching. Globalstar and OneWeb do not intersatellite switch, Iridium, LEOSAT, and SpaceX satellite switch either with RF links (K-band links for Iridium) or with optical transceivers (LeoSat and SpaceX).

The OneWeb example in their FCC filing stated that it would need at least 50 gateways. Gateways can cost $50 million to build or to move. The Australian government has asked Inmarsat to move its Perth Earth station, now

Table 6.4

Antennas: A New Earth Station in Motion (ESIM) and Stationary Earth Station (SES) Model

Passive Flat			Passive Conformal: Bespoke			Active Flat			Active Conformal: Bespoke		
256	512	1,026	256	512	1,026	256	512	1,026	256	512	1,026
Lowest cost		Low cost	Lowest cost		Low cost	High cost		Highest cost	High cost		Highest cost
Heat-insensitive			Heat-insensitive			–50°C to +125°C			–50°C to +125°C		
Resolves LEO to MEO to GSO interference						Resolves LEO to MEO to GSO interference					
Potential scaling from VHF to E-band: 138 MHz–95 GHz, including 5G bands from 450 MHz to 3.8 GHz						Potential scaling from VHF to E-band: 138 MHz–95 GHz, including 5G bands from 450 MHz to 3.8 GHz					
Vertical down only						Horizon-to-horizon multi-SAT LEO, MEO, GSO mixed constellation performance					
Low-cost consumer			Low-cost IOFMO, IOSMO, IOSO			High-value IOFMO, IOSMO, IOSO					
						Resolves issue of fast rate of rate decline					
Enables 5G in band cosharing?						Enables 5G in band cosharing?					

surrounded by high added-value housing, to a purpose-built space park within the next 5 years, which understandably Inmarsat would sooner not do.

Figure 6.15 shows a ground base station today with dishes pointing at different orbital slots.

Earth gateways in the future could be implemented as flat panel arrays with high-count antenna elements that could deliver substantial gain and interference mitigation. They could be integrated into large solar panel arrays. These panels could also function as a very large aperture active electronically steerable array with multiple adaptive beams supporting GSO, MEO, and LEO satellites. This would allow higher numbers of operators to coshare these expensive but critical ground assets.

6.18 Interconstellation Switching: GSO Satellites as the Mother Ship and the GSO as a Space-Based Server

The other way in which the number of Earth stations can be reduced is for LEO and MEO satellites to send their feeder links up to a GSO and then down to Earth to a GSO Earth station. This is already used, for example, by the Hubble telescope and the International Space Station (see Chapter 7). GSO operators could therefore be relatively relaxed about LEO and MEO operators directly connecting to consumers and business users on the basis that the traffic will then pass over the GSO network. However, routing via a GSO does introduce additional end-to-end latency.

Figure 6.15 A ground Earth gateway station today. (With thanks to Inmarsat.)

Note that in the future a single operator could potentially own or have access to all five delivery platforms: GSO, MEO, LEO, 5G backhaul, and 5G base station to user links. This would enable optimized routing to be implemented.

GSO satellites are generally a good place to put server bandwidth. Their power comes for free and they are well positioned to store and process information captured from smartphones, IoT devices, cars, trucks, and planes with data routed directly upwards via LEO or MEO satellite.

6.19 Upwardly Mobile Interconstellation Switching as a Way of Reducing the Number and Cost of Earth Stations

In many cases, it may be possible to reduce control signaling on gateway telemetry uplinks and downlinks through a combination of intersatellite and interconstellation switching. Irrespective of whether this is realized at RF (K-band) or optical frequencies, intersatellite switching and interconstellation switching allows satellites to work out their orbit position relative to all other satellites in their own and adjacent orbital planes. Autonomous station keeping is already proposed for CubeSats but could be more generally adopted. This would reduce the signaling bandwidth traffic from Earth station gateways and potentially reduce the complexity and number of those gateways.

6.20 Flat VSATs on Satellites

A delegate at one of our workshops in Singapore made the casual but profound comment that the satellite industry needed to focus its attention on ground-based rather than space-based innovation. The previous section on active and passive flat panel arrays hopefully demonstrates the truth of that simple statement.

However, replacing fixed spot beam antennas by adaptive flat panel arrays on space systems potentially provides increased beam-forming flexibility including the ability to support cell diameters down to 2 km or less. High beam count satellites with narrow beamwidth antennas allow individual users and IoT devices to be supported within those cells. This ability to scale from small cell diameters to country-wide and continent-wide beams is a unique capability available from satellite constellations. Note that coverage footprints can be shaped to follow the borders of countries or parts of countries or continents or parts of continents or oceans and parts of oceans or land-locked lakes, important, for example, in Brazil. This means that beam forming can be responsive to geographically specific marketing campaigns. From a link budget perspective, it also maximizes available flux density over the targeted coverage area.

6.21 Summary

In the last three chapters, we have covered launch technology innovation, satellite technology innovation, and antenna innovation. Antenna innovation is beneficial to all radio networks, terrestrial and nonterrestrial. In the context of satellite delivery economics, the specific benefits are that cell sizes (radius) can be scaled from 2 to more than 2,000 km to deliver geographic and demographic bandwidth on demand at a sufficient flux density on the downlink and sensitivity on the uplink to support mobile and fixed broadband connectivity.

Crucially, the active antennas and the signal processing algorithms that have been developed initially for military and more recently automotive radar are being repurposed into terrestrial and satellite communication. Delivering these products at consumer price points remains a challenge but the ability to embed a 16 or 32-element antenna into a smartphone screen or wireless wearable would unlock a 4-billion-unit market opportunity.

An interesting capability of these antenna systems is that they can calculate the angle of arrival and signal strength of both wanted and unwanted RF signal energy. This means that they can also calculate the required angle and power needed for the return path.

This is a critical part of the narrative that we hope is becoming apparent at this point in the book. There are many emerging opportunities to separate multiple radio systems in terms of the angle of arrival of wanted and unwanted signal energy. This includes the potential capability to reuse spectrum between users separated in three dimensions, for example, enabling high-count LEO constellations to coshare spectrum with MEO and GSO constellations and with 5G terrestrial networks including backhaul.

Passive and active flat panel and conformal antennas, which we have called flat VSATs, also help to resolve feeder link interference issues, particularly in K-band, and potentially reduce the cost of Earth gateways.

Flat VSATs therefore resolve many of the potential tension points between GSO, MEO, and high-count LEO constellations and can deliver major cost savings both in terms of Earth and space asset capital cost and operational cost. Flat VSATs open the opportunity for the satellite to capture some of the connectivity value of smartphones and the emerging wireless wearables sector based on low element count antennas built into smartphone display screens. Larger footprint applications can scale the element count to provide more precisely defined angular resolution to support very high data rate (multiterabit) interconnections to and from optimized LEO, MEO, and GSO space platforms.

Satellite networks have an evolving role in helping terrestrial networks to meet their 5G energy efficiency and carbon footprint targets both in terms of backhaul power consumption and base station and user device IoT power drain.

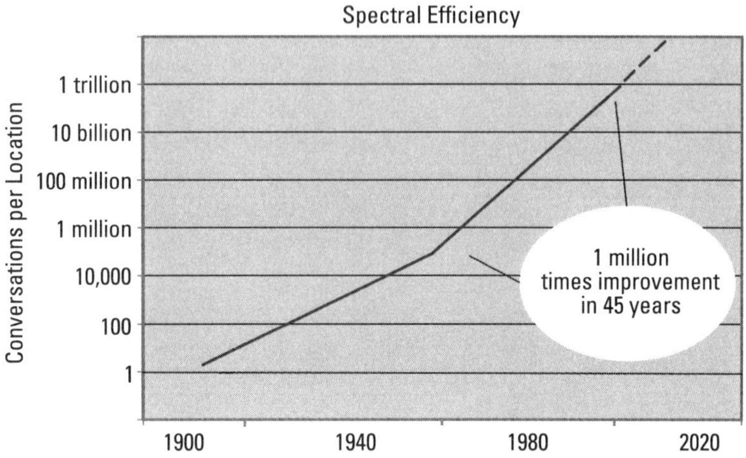

Figure 6.16 Cooper's Law of spectral efficiency.

Antennas are a critical part of this story with the challenge of producing adaptive electronically steerable antenna arrays (AESA) or passive arrays that can achieve efficient and effective angular power separation at consumer price points.

Twenty years ago, Marty Cooper, the man credited with developing the first commercial cellular phone [24], suggested that spatial separation would prove to be an important aspect of terrestrial cellular system design and one of the ways in which spectral efficiency could be improved.

The eponymous Cooper's Law of spectral efficiency states that the maximum number of voice conversations or equivalent data transactions that can be conducted in all of the useful radio spectrum over a given area doubles every 30 months (Figure 6.16).

His company, Arraycomm, produced many of the initial processing algorithms that have been subsumed into present-day MIMO and AESA systems.

However, the story is not just about new antenna materials and manufacturing techniques or just about terrestrial spatial processing and angular power separation but about integrating that innovation with constellation innovation and a three-dimensional (3-D) model that allows terrestrial 5G to coshare with LEO, MEO, and GSO radio systems.

References

[1] http://ethw.org/Smith_Chart.

[2] https://5g.co.uk/guides/what-is-lte-broadcast/.

[3] 3GPP TSG RAN meeting#75, RP-170132.

[4] http://www.etsi.org/technologies-clusters/technologies/satellite/satellite-umts-imt-2000.

[5] http://www.ic.gc.ca/eic/site/smt-gst.nsf/eng/h_sf09857.html.

[6] https://www.itu.int/rec/R-REC-P.681/en.

[7] http://www.radio-electronics.com/info/antennas/waveguide/rf-waveguide-dimensions-sizes.php.

[8] https://www.everythingrf.com/tech-resources/waveguides-sizes.

[9] http://www.pennengineering.com/e-and-h.php

[10] http://miwv.com/images/Waveguide-Chart.pdf.

[11] http://www.rfcom.co.uk/index.php.

[12] http://www.photonicsinc.com/index.html.

[13] Dr. Amitabha Ghosh, IEEE webinar, September 20, 2017.

[14] http://www.arraycomm.com/.

[15] http://www.quintelsolutions.com/.

[16] http://www.antenna-theory.com/antennas/patches/pifa.php.

[17] http://www.mrs.org/58th-emc-topics/wide-bandgap-materials.

[18] https://www.metawave.co/.

[19] https://www.kymetacorp.com/markets/connected-car/connected-car/.

[20] http://www.phasorsolutions.com/.

[21] http://www.anokiwave.com/company/company-news/releases/awmf_0108.html.

[22] http://plasmaantennas.com/technology/psian-plasma-antennas/.

[23] https://www.fcc.gov/general/specific-absorption-rate-sar-cellular-telephones.

[24] http://www.destination-innovation.com/how-startrek-inspired-an-innovation-your-cell-phone/.

7

Constellation Innovation

7.1 Technical and Commercial Factors Determining and Driving Constellation Innovation

We ended Chapter 6 by stating that a combination of technical and commercial factors are reducing the per bit delivery costs of the satellite industry. Technical factors include launch innovation, satellite innovation, and antenna innovation. Commercial factors include lower insurance costs, a by-product of improved launch reliability and longer in-orbit service life, multiple payload protocols, and, most significantly, an infusion of cash and equity from companies such as Google, Apple, Facebook, and Amazon. Twenty years ago, these companies did not exist (Facebook), were being started (Google), were less than 4 years old (Amazon), or were only just emerging from a period of losses (Apple).

In this chapter, we cover companies that are either newly formed or that are at most 4 to 5 years old with ambitions to build satellite constellations at a scale previously attempted (Teledesic and Skybridge) but never achieved.

The scale of these constellations is important because this determines their economic viability but, as covered in previous chapters, high-count constellations also mean that satellites are nearly always nearly overhead. This minimizes latency, improves the link budget, and makes it more likely that connections with terrestrial users and devices will be clear line of sight, minimizing scatter loss and surface absorption. This is particularly important for the higher-end centimeter bands and millimeter bands. High-count constellations can also take advantage of the capabilities of active electronically steerable (AES) arrays on terrestrial platforms such as cars and trucks. Advanced implementations of

AES arrays are capable of horizon-to-horizon scanning for satellites providing the most favorable link budget. For example, if a preferred signal from a directly overhead satellite is blocked, then a satellite lower in the sky can provide alternative connectivity.

If coverage into a building is needed, this is more likely to be achieved via a lower elevation satellite through the window, although the link budget and latency will be less than optimum. The better option is to have a flat panel array on the roof pointing directly upwards with Wi-Fi into the building.

7.2 The Point of Constellation Innovation

The point of constellation innovation can be summarized as: make satellites do more for longer at lower cost. This includes techniques for increasing service life, for example, through improved hardware (processors, memory, and solar panel arrays), in-orbit servicing, hardware upgrades and repairs, and optimized station keeping. Low Earth orbit (LEO) satellites, particularly lower-orbit LEO satellites, have a small amount of atmospheric drag and can be pulled off station by changes in the near-Earth gravitational field. Station keeping is therefore a continuous task that absorbs power and, in the past, has been an important factor determining end of life, for instance, due to expended hydrazine.

Solar panel ion thrusters have helped but also optimized ground based control of satellites both in terms of their altitude and attitude (pitch and yaw control) have resulted in an extended service life. For example, the first generation of Iridium satellites launched in the 1990s had an expected life expectation of 7 years but the constellation remained fully functional for 20 years and is only now being replaced. It has also been traditional practice to have a core of operational satellites with some backup satellites that are either kept in an interim orbit or kept ready for rapid launch. Remember that it only takes 20 minutes to get into space, faster than driving a truck to a base station site in rush hour, although more fuel will be used on the journey.

More recently, constellation design has embraced the concept of autonomous self-drive satellites that manage their own station keeping independently of any Earth-based network control. This theoretically at least reduces earth to space signaling overhead and could be potentially more power-efficient. CubeSats over their 15-year implementation history (the first CubeSat was launched in 2003) have moved towards this model of autonomous or semiautonomous control, although there are related regulatory issues such as debris limitation and avoidance that need to be accommodated [1]. CubeSats with optical (or RF) intersatellite switching can continuously calculate their relative separation distance and use that information to do station keeping without reference to a terrestrial control function.

7.3 A Reminder of the Constellation Options

First, we have a reminder of the constellations options including the spectrum options (excluding very high frequency [VHF]) (Table 7.1), orbit options (excluding Quazi zenith and high elliptical orbits) (Table 7.2), and size (including CubeSats) (Table 7.3).

In terms of commercial constellations and excluding for the moment CubeSats that will be covered later, there are four distinct sectors that need to be considered, as listed in the following sections.

7.4 NEWLEGACYLEO

NEWLEGACYLEO operators are companies such as Orbcomm (VHF), Iridium (L-band), and Globalstar (L-band and S-band) that have either upgraded their constellations (Orbcomm OG2 and Globalstar) or are in the process of upgrading (at the time of this writing, Iridium had successfully launched 30 Iridium NEXT satellites on three SpaceX rockets). Services include IoT and voice connectivity and positioning, location, and collision avoidance. These constellations are in polar orbits and provide coverage at all latitudes including obviously polar regions where GSO to terrestrial connectivity becomes problematic.

7.5 NEWLEGACYGSO

NEWLEGACYGSO are companies such as Inmarsat and Intelsat and all the other SATs with geostationary satellites providing a range of broadcasting and two-way data and voice services. Note that geostationary is not the same as geosynchronous. Geosynchronous satellites are satellites that are rotating at the same speed as the Earth but are not positioned over the equator. They are sometimes described as quasi zenith (QZ) constellations and will usually produce a defined parabolic footprint over a defined geographic area. The new Mitchibiki GNSS constellation in Japan is an example.

These constellations are significantly different in terms of their spectrum and available bandwidth. The Orbcomm VHF constellation has a 1 MHz + 1

Table 7.1
Satellite Spectrum Options

L-Band	S-Band	C-Band	X-Band	Ku-Band	K-Band	Ka-Band	V-Band	W-Band
GHz	GHz	GHz	GHz	GHz	GHz	GHz	GHz	GHz
1–2	2–4	4–8	8–12	12–18	18–27	27–40	40–75	75–110

Table 7.2
Orbit Options

LEOs	Medium Earth Orbits	Geostationary Orbits
160–1,200 km	8,000 km (O3b)–20,000 km (GPS)	36,000 km

Table 7.3
Satellite Sizes

Pico	Nano	Micro	Macro
1 kg	19 kg	<500 kg	>500 kg

MHz passband at VHF, and Iridium and Globalstar have a 10 MHz + 10 MHz passband with an additional 7 MHz for Globalstar in S-band. The NEWLEGACYGSO high throughput satellites have 3.5-GHz passbands at Ku-band, K-band, and Ka-band combined with L-band, S-band, and C-band transponders. The satellites scale from 170 kg for an Orbcomm LEO satellite to 6,000 kg for an Inmarsat or Intelsat GSO, which means that they scale from tens of watts to hundreds of watts to kilowatts in terms of available power.

There are opportunities to exploit these differences to combine the performance benefits of each option, for example, a robust link budget with Orbcomm but limited bandwidth but with other capabilities such as positioning and location and good coverage at high latitudes. Similar functionality is available from Iridium and Globalstar though with additional bandwidth. The NEWLEGACY GSOs with their 15 kW of solar power and 3.5 GHz + 3.5 GHz of Ka-band spectrum hold the high throughout trump cards but have the longest latency, long path lengths through the atmosphere at higher latitudes and a vulnerability to building and foliage blocking at higher latitudes.

Sometimes launch failures or partial failures where satellites fail to reach their final orbit or lose RF power prompt interworking agreements. Orbcomm, for example, has had a number of satellite failures as part of their OG2 constellation upgrade with almost one-third of the constellation (10 of 31 satellites) compromised either by a failure to achieve final orbit or hardware or software failures. At a book value of $10 million per satellite, this is a frustration, although not unknown or even particularly unusual in the satellite industry [2]. The insurance was for launch and 1 year in orbit and therefore the satellites with hardware and software failure must be written off. The technical and commercial solution has been to couple the service proposition to Inmarsat's I-4 GSO L-band service offer and to reposition the surviving fully operational OG2 satellites to optimize high-latitude coverage. The decision was also taken to develop modems that could combine OG2 service with Inmarsat 4 (L-band)

and terrestrial cellular, initially AT&T, Verizon, T-Mobile, Orange, Telefonica, Vodafone, and Rogers (Canada).

The remaining OG2 satellites produce a longer gap than originally intended between passes, which results in connection latencies of several minutes (which is often not a problem for IoT connectivity). At higher latitudes, users connect with the Inmarsat I-4 with a latency of around 15 seconds, a consequence of available bandwidth rather than visibility constraints. Note that fairly obviously a GSO constellation can have visibility at higher latitudes if users are at high altitude. From a customer's point of view, the end result will be similar to the original constellation plan.

As covered in Chapter 2, Orbcomm customers include vendors of high-value, large mobile machines, monster tractors and mining machines, diggers, and generally devices that dig very large holes in the ground and a few ships and oil and gas rigs as well with enough capacity now available from the combined LEO and GSO and cellular service offer to move into more mass market telematics [3]. A potentially catastrophic combination of launch failure and hardware and software failure in space has had a minimal impact on Orbcomm customers and little obvious impact on the growth and profitability of Orbcomm, which proves the point that a mixed constellation approach combined with terrestrial cellular is probably the way ahead for many space and terrestrial service providers. In fiscal 2016/2017, Orbcomm generated $57 million in revenue, a 13.8% year-over-year increase, and gained 62,000 net subscribers. The company's total billable subscriber count reached 1.83 million, up 10.8% from the previous year's 1.65 million.

Inevitably, scale constraints associated with developing and manufacturing modems for specific network combinations imply additional cost particularly as the satellite networks have different physical layers in different bands requiring bespoke RF hardware (PA, filter and switch paths, antennas and matching networks). This does not matter for monster machines costing large amounts of money. It does matter for low-cost IoT connectivity.

The Orbcomm constellation also offers an Automatic Identification Service (AIS) for large ships. The objective is to stop large ships hitting each other. The system latency is claimed to be of the order of a minute.

7.6 NEWLEO

The NEWLEO constellations are being created or promoted by companies such as OneWeb, SpaceX, and LeoSat and in the CubeSat sector, companies such as Sky and Space Global. Last but not least, there are medium Earth orbit (MEO)/GSO constellations with SES/O3B being the main player in this sector.

Table 7.4 lists these four sectors and the principal participants in each sector, the number of satellites either planned or in orbit, the spectrum allocation, passband, satellite size/weight, typical data rate, and functionality. This is a fast-changing industry so this table represents a snapshot in time rather than a long term record of constellation availability and gives examples rather than a full list of the companies involved.

Table 7.5 should be interpreted with similar caution but provides a summary taken from U.S. Federal Communication Commission (FCC) filings extant in 2016 and 2017 for new LEO constellations including their spectrum requirements, satellite count, orbit altitude, and weight.

The total satellite count in space today is somewhere around about 4,000 satellites and Table 7.5 lists proposals to increase this twice over. This implies a rigorous regime of space debris management, a topic that we address at the end of this chapter.

7.7 NEWLEGACYLEO

7.7.1 Iridium

Note that these NEWLEOS compete with existing LEOS that have twenty years of space experience and established and profitable and well-established presence in a broad cross-section of vertical markets and customer sectors.

We have already referenced the ongoing NEXT upgrade of the Iridium constellation (30 satellites launched in the first 10 months of 2017 via three SpaceX rockets) and highlighted that the new generation constellation does more than just provide voice and data services and now includes positioning and location exploiting the strong Doppler signature of the constellation (and higher flux density than equivalent GNSS MEO constellations).

Twenty years of playing in space has also produced an established and stable vendor supply chain. Companies involved in the NEXT constellation upgrade include Harris, Hughes, Honeywell, Boeing, SpaceX, Thales Alenia, and Cambridge Consultants. Note that this highlights a substantive difference between the satellite industry and mobile broadband terrestrial mobile operator community. Mobile operators essentially have three vendors (Huawei, Ericsson, and Nokia) that provide them with most of the RF plumbing needed to deliver 4G and 5G services. Some but not all operators also have extensive in-house research and development and engineering and implementation support teams, although this is becoming less common. Satellite operators generally have a much more diverse supply chain and are more reliant on outsourcing. SES, for example, has a grand total of 69 employees with a turnover per employee of $33 million.

Table 7.4
Four Satellite Sectors and Examples of Companies Active in Those Sectors

	NEWLEGACYLEO			NEWLEGACYGSO		NEWLEO		MEO/GSO
	Orbcomm	**Iridium**	**Globalstar**	**Inmarsat**	**Intelsat**	**OneWeb**	**SpaceX**	**SES/O3b**
Orbit (km)	775	780	1,410	36,000	36,000	1,200	625	8,062
Number of Satellites	31	66	24	12	20	2,650	4,000	20
Spectrum	VHF	L-band	L-band and S-band	L-band, S-band, K-band and Ka-band	C-band, Ku-band, K-band, Ka-band	Ku + Ka-band	Ka-band	K-band/Ka-band
Allocation	137–138 MHz, 148–149 MHz	1,616–1,626 MHz	1,610–1,618, 2,483–2,500 MHz	17.7–21.2 GHz, 27.5–31 GHz	17.7–21.2 GHz, 27.5–31 GHz	Ku-band, 12.2–12.7 GHz	17.7–21.2 GHz, 27.5–31 GHz	17.7–21.2 GHz, 27.5–31 GHz
Passband	1 + 1 MHz	10 MHz	10 + 17 MHz	3.5 GHz	3.5 GHz	500 MHz + Ka-band	3.5 GHz	3.5 GHz
Weight	170 kg	860 kg	700 kg	6000 kg	6,000 kg	125 kg	125 kg?	700 kg
RF Power	250W	200W	200W	15 kW	15 kW	200W?	200W	2.4–1.7 kW
Upgrade	OG2	Next	Upgrading	Upgrading	Upgrading	2018 launch	2018 launch	Upgrading
Data Rates	Bps/kbps	kbps	kbps	Mbps	Mbps	Mbps	Mbps	Mbps

Table 7.5
NEWLEO Constellation Proposals Taken from 2016/2017 FCC Filings

Iridium	Globalstar	Sky Space Global	OneWeb	SpaceX	LeoSat	Boeing
L-band	L-band and S-band	UHF, L-band and S-band	Ku-band and Ka-band	Ku-band and Ka-band	Ka-band	V-band and C-band
78 LEOS	24 LEOs	200 LEOs	650 LEOs	4,000 LEOs	78 LEOs	2956 LEOs
780 km	1,414 km	500–800 km	1,200 km	1,200 km	700 km	1,200 km
860 kg	700 kg	10 kg, CubeSat	200 kg	100–500 kg	860 kg	?

This may change and indeed is probably already changing and it might be expected that the two supply chains might become more similar over time. In particular, satellite operators with ambitions to deliver mass-market services at consumer price points will need a supply chain that is geared to the scale economics that will be needed. We return to this topic in Chapter 8. There will also need to be consumer marketing teams to work in parallel with the government sector and business-to-business market and sales teams.

The Iridium NEXT constellation reuses the same orbit as the previous constellation and has the same L-band uplink and downlink spectrum (see Chapter 2).

7.7.2 Globalstar

Originally formed in 1991 as a joint venture between Loral Systems and QUALCOMM, Globalstar has had technical and commercial challenges including RF hardware issues on some satellites and a period in Chapter 11, but has now completed an upgrade of the original constellation with 24 new satellites in three orbital planes. Globalstar shares the same L-band spectrum as Iridium but has a smaller satellite count (24 rather than 66 operational satellites), which means lower capacity and fewer satellites visible at any time (two satellites visible at any one time in temperate zones).

The satellite is bent pipe architecture and does not intersatellite switch, which means that more Earth stations are needed. This adds OPEX and CAPEX cost to the constellation but also introduces additional challenges managing potential interference between the uplink and downlink feeder links and other spectrally and geographically proximate Earth and space-based radio systems (see Chapter 2 for more details on this). Globalstar claimed that bent pipe rather than regenerative architecture (where the uplink is demodulated and then modulated and coded again for the downlink) produces a lower latency, although in practice this is dependent on the link budget.

The top of the Globalstar lower C-band allocation (5,091–5,250 MHz) is immediately adjacent to the 5-GHz Wi-Fi band from 5,250 MHz to 5,925 MHz including 8.2.11p (for automotive connectivity) at the top end of the band (5,825–5,925 MHz).

The new satellites have a life expectation of 15 years (twice the life of the previous generation) and support a data speed of 256 kbps (up from 9.6 kbps from the first generation constellation). The first constellation had a build out-cost of $5 billion, and the second constellation 20 years later cost $1 billion.

In addition to the L-band user links, Globalstar has 11.5 MHz of S-band spectrum allocated under FCC Part 25 rules providing full interference protection from adjacent services. The band is immediately above the intensively occupied 2.4-GHz Wi-Fi band.

The band is immediately below FDD Band 7 and TDD Band 41 (Table 7.6). The proposed terrestrial physical layer is TD-LTE.

This is promoted as a global band, but, in practice, many countries have deployed FDD Band 7 with TDD spectrum in the duplex spacing between the upper and lower passband.

Although this might seem like a golden opportunity to roll out a global hybrid LTE and satellite phone network, it may, in practice, be frustrated by regional and country-specific implementations of LTE, which are not compatible with the Globalstar TDD-LTE ground component.

Note also that, to date, this is a U.S. FCC rather than global spectrum assignment, so Globalstar has to address regulatory approval in rest of the world markets or gain ITU acceptance of the subband as a global LTE band.

Similarly, while this might also seem like a golden opportunity to integrate the network with 2.4-GHz and 5-GHz Wi-Fi, there could be potential out-of-band interference issues with 2.4-GHz Wi-Fi and in-band interference issues with 5-GHz Wi-Fi, which could frustrate commercial exploitation.

Table 7.6
Band 7 and Band 41

Globalstar S-Band	**Band 7**		
	Mob TX FDD	Mob TX/RX TDD	Mob RX FDD
2,483.5–2,500 MHz	2,500–2,570 MHz (70 MHz)	2,570–2,620 (50 MHz)	2,620–2,690 MHz (70 MHz)
	Band 41		
	2,496	190 MHz including guard band	2,690

7.7.3 Device Availability for Hybrid Cellular/Satellite Constellations

With QUALCOMM's long-term links with Globalstar, it might be expected that a large range of market competitive devices could be made available.

Devices listed at the time of this writing include a mobile phone with satellite functionality (see Figure 7.1), a SAT-Fi mode providing a Wi-Fi hot spot coupled to the satellite and the TDD LTE physical layer, and a range of trekking, leisure, and IoT devices.

What is really needed is satellite connectivity to be added to smartphones as standard. This is unlikely to happen due to the lack of global market scale (the Band 7/Band 41 issue). It would be theoretically possible to increase the passband to include the 11.5 MHz of Globalstar spectrum, but the wider passband would decrease the sensitivity and selectivity of the Band 7 and or Band 41 LTE RF switch path and add to the cost of the phone. This is unlikely to be acceptable to the major handset vendors or their operator customers.

Inmarsat are likely to have similar issues with their European Aviation Network (EAN), although in their case their spectrum is adjacent to Band 1. As a reminder, Table 7.7 shows the EAN band plan (last seen in Chapter 2).

7.8 NEWLEO Angular Power Separation

As already discussed, the NEWLEOs are significantly different to the legacy LEOs both in terms of their satellite count and available bandwidth and orbit RF power but also in the way that they propose to share spectrum with the MEO and GSO satellites flying above them.

This brings us back to the knotty topic of progressive pitch and angular power separation. In Chapter 2, we referenced Skybridge and Teledesic as

Figure 7.1 Globalstar user device.

Table 7.7
Another S-Band Adjacency Challenge/Opportunity

Band 1	Inmarsat S-Band	Band 1	Inmarsat S-Band
Mob TX	Earth to space	Mob RX	Space to Earth
1,920–1,980	1,980–2,010	2,110–2,170	2,170–2,200

the two entities that first introduced the concept of progressive pitch angular power separation as a mechanism for cosharing spectrum with MEO and GSO constellations. This was 20 years ago, but the basic principle remains the same, although with substantial fine-tuning now possible. Put simply, a LEO satellite at high latitude can look directly downwards at a LEO satellite dish or flat panel array antenna on Earth pointing upwards. A dish at the same location looking at a GSO satellite will be focused at a low elevation angle pointing close to the horizon, which should mean that it receives no signal energy or not much signal energy from the overhead LEO.

At the equator, both the GSO and LEO (and MEO) satellites will all be looking directly downwards and the satellite dishes will be pointing directly upwards. If the same operator in the same spectrum manages all three constellations, then you could have a scenario in which equatorial users could uplink and downlink simultaneously from all three constellations, which would produce some interesting and useful amounts of Ku-band, K-band, and Ka-band width combined with interesting and useful amounts of downlink power and uplink sensitivity.

However, if different operators manage different constellations, there is an inherent interference issue. Established GSO operators, for example, that are currently investing in high throughput GSO satellites in Ku-band, K-band, and Ka-band are less than enthusiastic about a NEWLEO operator pouring unwanted signal energy into their GSO terrestrial receive passband.

The solution promoted by the NEWLEOs is to switch off transmissions as the satellites pass over the equator. As the NEWLEO satellite moves away from the equator, it can restart transmission but needs to ensure that the elevation angle is sufficiently acute to avoid interference with GSO Earth-based receivers. This is accomplished by changing the pitch of the satellite progressively, hence the description progressive pitch but combined with power control and handover to other satellites operating at a lower elevation angle.

The detail of how this is done is dependent on orbit altitude, orbit speed/duration (a consequence of altitude), orbit path, RF power and sensitivity, and regulatory constraints (agreed protection ratios).

For the NEWLEOs, the best link is always going to be directly overhead, so any lower angle will require a longer path through the atmosphere and a higher chance of blocking from buildings and foliage. Latency will also increase and the link budget will be lower and more variable. Because the signal needs to pass through more atmosphere, it is more likely to be affected by rain fade. Agreeing coexistence conditions and ratifying these agreements through legal and regulatory channels are therefore critical to the economic viability of these new constellations.

Figure 7.2 shows the principle of angular power separation applied to the coexistence of the O3b MEO network and SES GSO. Given that SES now fully owns O3b, it might be thought that this would be relatively noncontentious, but any and all other GSO operators sharing this spectrum also need to be reassured that protection ratios can be managed and any in-band or out-of-band interference limited to a level where no economic cost is projected on other operators or alternatively, if there is an economic cost, that it is fully compensated. There are also adjacency issues with military users that need to be accommodated (see Chapter 2).

Figure 7.3 shows the potential complexity of this calculation in terms of the satellite count of both constellations. Figure 7.4 shows the number of ground stations and gateways and the coverage map.

7.9 OneWeb Coexistence

The issues around MEO/GSO coexistence are a mere warm-up act for the debate that is presently swirling around NEWLEO interference calculations with

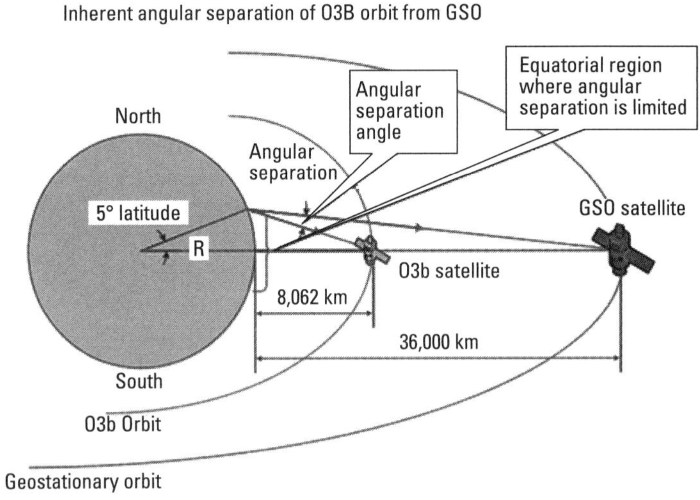

Figure 7.2 Angular separation of the O3b MEO constellation from the SES GSO constellation. (Thanks to SES networks.)

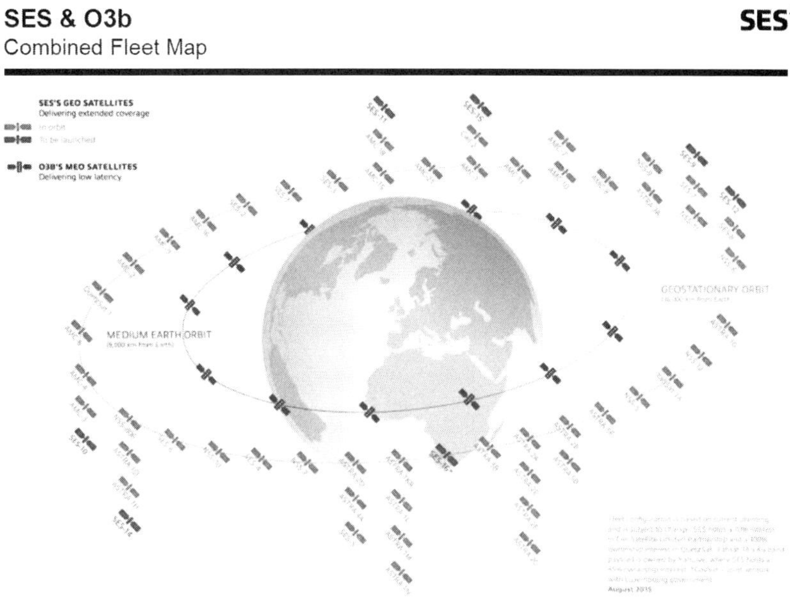

Figure 7.3 SES and O3b fleet map. (Thanks to SES networks.)

OneWeb as the poster child of the sector. First, we will discuss OneWeb's network topology. More detail is available from the FCC Web site [4].

7.9.1 OneWeb Earth Stations

OneWeb proposes to have 50 gateway Earth stations with at least four deployed in the United States including Hawaii and Alaska. The gateway Earth stations will also transmit and receive control channels for satellite payload control and gateway link power control. A subset of gateway sites in high-latitude regions of the world will provide telemetry tracking and command (TT and C). There will be at least two separate satellite control centers, probably in Virginia, United States, and the United Kingdom, with network operations controlled from the United Kingdom and Melbourne, Florida. Payload control transmissions to and from the U.S. gateway Earth stations will take place in the band edges just below 19.3 GHz (downlink) and just above 27.5 GHz (uplink).

7.9.2 OneWeb Progressive Pitch

Each OneWeb satellite will have 16 nominally identical user beams, operating in Ku-band, each consisting of a nonsteerable highly elliptical spot beam. There are also two identical steerable gateway beam antennas operating in Ka-

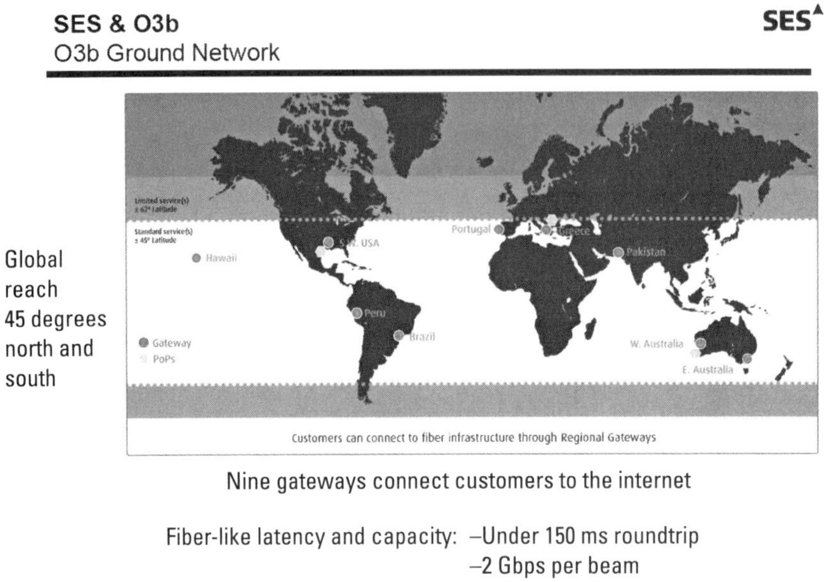

Figure 7.4 O3b ground stations, gateways, and coverage map. (Thanks to SES networks.)

band. Each of these antennas creates an independently steerable circular spot beam. The second beam tracks the next gateway Earth station for handover procedures.

While the Ku-band user beams cannot be steered, the Ku-band footprint can be moved up and down in latitude. The attitude control system of each satellite allows the pointing of the satellite to be adjusted so the beam pattern can be moved in the pitch direction (north to south), hence the term progressive pitch. The power output from each beam can also be controlled and adjusted to meet EIRP and flux density country-specific regulatory requirements.

The movement of the satellites in their orbits means that a user will be progressively handed over from beam to beam within a OneWeb satellite and then handed off to the beams on the next satellite in the same orbital plane or adjacent plane.

Each user beam supports services to multiple user terminals. In the forward direction (gateway to user), there is a TDM transmission scheme within a single 250-MHz channel. Each user terminal in the beam receives and demodulates the whole carrier and extracts only the data destined for it which is determined by the data headers (and potentially also the position in the TDM transmission). In the return direction (user to gateway), there is a single carrier TDMA/FDMA transmission scheme modulated on to a relatively narrowband carrier (1.25 MHz to 20 MHz wide).

Based on 720 satellites, every point on the Earth's surface will always see a OneWeb satellite at an elevation of no less than 55° with increasing minimum elevation with latitude.

7.9.3 OneWeb Interference Models

The FCC power flux density (PFD) requirements and associated calculated allowable EIRP are documented in detail in the FCC and ITU filings. The ITU PFD limits applicable to NGSO systems in the 10.7–11.7-GHz band can be found in Table 21.5 of the ITU radio regulations and are effectively the same as the FCC PFD limits. OneWeb document their methodology for calculating maximum EIRP density for all angles of arrival and departure across the Ku-band, K-band, and Ka-band passbands taking into account the spreading loss from the satellite to the surface of the Earth (i.e., the variation in signal strength over the elliptical area illuminated). They review their methodology for demonstrating compliance with the FCC and ITU EIRP and PFD limits established to minimize interference to GSO systems and their methods for calculating interference from GSO satellites into the OneWeb network including other mechanisms such as power control, which are used to deliver constant flux density at different elevation angles. Note that lower elevation angles will absorb more RF power and will therefore reduce the capacity of the constellation, which, in turn, reduces the technical and commercial efficiency of the constellation. These technical details and the interference and coexistence assumptions behind these technical details determine the economic viability of the business model.

Progressive pitch, also described by OneWeb as pitch bias, takes place gradually as the satellite passes through mid-latitudes to lower latitudes. As the satellites pass over the equator, they temporarily turn off RF power and then adjust their pitch to the opposite direction. Pitch adjustment is managed by reaction wheels.

OneWeb documents its methodology for demonstrating that it complies with the required protection ratios for Ka-band GSO networks. Note that the gateway antennas on the OneWeb satellites are fractional beamwidth (less than half a degree half-power bandwidth) in order to minimize the number of required gateway sites. This yields a gain of typically 55 dBi on the transmit path and 51.5 dBi on the receive path.

OneWeb is required to demonstrate that the siting of its gateway sites and the constraints on the number of possible positions of OneWeb satellites with which each gateway site can communicate will reduce OneWeb gateway to GSO Ka-band interference to an acceptable level. OneWeb is proposing to operate its user terminal Earth stations on a noninterference nonprotected basis, which means that the receiving Earth stations will not seek interference protection from fixed services in the band.

There are also fixed service links in the 12.2–12.7-GHz band issued under old FCC allocations. These are now grandfathered, which means that no new allocations will be made but OneWeb is still required to demonstrate that the protection ratios for these services are met.

7.9.4 OneWeb Coexistence with GSO Systems

The filing document addresses coexistence with the Multichannel Video and Data Distribution Service (MVDDS) with proposed sharing mechanisms based on database sharing of each system's transmitters and receivers in the 12.2–12.7-GHz band (shades of sub-1-GHz white space). Dish Networks deliver TV over these bands and are mounting a rigorous defence of MVDDS spectrum access rights in the United States.

Interference mitigation with respect to terrestrial networks in the 17.8–18.3-GHz band is also addressed. The band is used by OneWeb in the space to earth direction for a relatively small number of Earth stations. Similar procedures apply with respect to terrestrial networks in the 27.5–28.35-GHz band. Terrestrial Local Multipoint Distribution Systems (LMDS) have access to this spectrum on a primary basis and are licensed by the FCC by geographic area. The position of OneWeb gateways will therefore need to be coordinated with any local LMDS operators. OneWeb as a secondary user in this band has to accept incoming interference from the primary user.

OneWeb will also need to coordinate with NASA to guarantee protection of Tracking and Data Relay Satellite Systems (TDRSS) in the 14.0–14.2-GHz band and with space observatories operating radio astronomy services in the 10.6–10.7-GHz band and with U.S. government satellite networks, both GSO and NGSO in the Ka-band.

Note that all these coexistence calculations need to account for the additive power received at victim receivers from all visible OneWeb satellites and Earth stations with the EIRP mask characterized for worst-case modulation and traffic patterns. This includes worst-case conditions defined as the maximum number of nongeostationary satellites transmitting and receiving simultaneously with overlapping frequencies from the associated Earth stations within a given cell defined on a per square kilometer basis. The calculations need to include the required minimum GSO avoidance angle.

Interference from a OneWeb satellite to a GSO also needs to take into account path length.

Demonstrating compliance with the coexistence conditions (protection ratios) for existing incumbent users in OneWeb's Ku-band, K-band, and Ka-band spectrum is complex and subject to legal challenge. This is just one market (the United States), one filing (the FCC), and one constellation and will need to be repeated for each and every other market in the rest of the world.

In particular, characterizing the mitigation achieved by angular power protection is open to technical interpretation and likely will be used by incumbents to defend their spectral access rights and market position. Note that the NEWLEO constellations have higher satellite count (hundreds or thousands of satellites) than the O3b MEO network [5] and many more satellites than any of the GSO operators (Intelsat has 40 orbital slots).

Dish Networks also have a patent pending on progressive pitch, as probably do other entities with NEWLEO constellation ambitions.

7.10 Angular Power Separation and Active Electronically Steerable Antenna Arrays

In Chapter 6, we covered AES arrays and their passive equivalents. AES arrays are particularly interesting when considered as an integral part of a NEWLEO with progressive pitch for several reasons.

AES arrays can do horizon-to-horizon scanning to evaluate the angle of arrival of wanted and unwanted signal energy and use this information to null out unwanted energy and provide gain to wanted energy received from a specific elevation angle and area of sky. If this data can be captured and consolidated, it gives a near real-time picture of both the serving network and other networks either cosharing the band or in adjacent spectrum.

On the transmit side, knowledge of angle of arrival allows an optimum angle of departure to be calculated minimizing uplink power consumption and minimizing interference into space based systems. AES antennas can perform these functions in any band from VHF to Q-band, although in practice size constraints make higher bands and shorter wavelengths a preferred option.

AES antennas are not going to be universal due to cost. Few TV subscribers would want to replace their low-cost Ku-band satellite dish with a planar array costing 10 times as much as the dish. Active arrays are also sensitive to large temperature gradients, for example, on the roof of a car or truck and therefore may not be an optimum choice for some applications.

This means that the NEWLEO constellations need to show that they will not cause interference into dish antennas or passive flat antennas.

7.11 Interference Calculations and Other Arguments

The problem with interference calculations is agreeing on how the calculations of interference should be done, particularly given that with high-count LEO constellations in particular there are no broadly based empirical measurements that can be used to validate and fine-tune propagation models, channel models, and statistically based interference modeling.

This is not a new issue for the industry. The TV industry spent 10 years questioning the assumptions and modeling methods used to calculate terrestrial LTE to TV interference in the 800-MHz band (first digital dividend), 700-MHz band (second digital dividend), and 600-MHz band (third digital dividend).

The default would have been for the mobile broadband operator community to accept a set of worst-case interference conditions, but this would have resulted in protection ratios that would have made much of the spectrum being auctioned either unusable or seriously compromised. The discussions also had to cover second-order effects such as the impact of pulse trains from the LTE signal compromising the automatic gain control of TV receivers. In the end, common sense prevailed, an amount of money changed hands, and everybody more or less lived happily ever after.

It could be expected that the same accommodation could be found between GSO satellite operators and the NEWLEO operators, but in practice the known unknowns and the unknown unknowns implicit in modeling interference theoretically without recourse to empirical data introduce an unsettling degree of uncertainty and room for debate and litigation.

7.12 Asia Broadcast Satellite Case Study

This is illustrated by a case study [6] undertaken by Asia Broadcast Satellite (ABS) [7] based on a presentation given at a cable and satellite broadcasting event in 2015 [8]. The case study looks specifically at the issue of OneWeb to TV interference in Ku-band.

Article 22 of the ITU Radio Regulations is referenced in terms of the definition of equivalent power flux density allowable from the NGSO (NEWLEO) constellation taking into account that the flux density (unwanted signal energy) could be the composite of all or some of the NEWLEO satellites visible from horizon to horizon. The assertion is made that compliance with this limit would not guarantee that satellite TV reception would not be compromised, and the study calculates the equivalent allowable effective isotropic radiated power from the OneWeb constellation assuming the Article 22 EPFD limits. The study then looks at the cone of visibility from the victim antenna and the pass rates of the satellites to calculate outage times. To complicate matters, the calculation then has to be done for multiple satellites. On the basis of the above assumptions, an interference model is proposed. The outputs of the model show that progressive pitch does not provide sufficient protection at low elevation and low altitude.

Self-evidently, OneWeb and or any other NEWLEO entity would use a different set of assumptions and end up with a completely different result, but the point we are trying to make is that this is a multidimensional modeling requirement that is in reality far more complex than the simple case study of NEWLEO to TV interference given that it needs to potentially comprehend multiple LEO constellations all with different orbits and numbers of satellites and RF power and interference mitigation techniques interacting with GSO and MEO satellites that are either in band or spectrally adjacent delivering angular power across an infinite range of inclination angles in a wide range of propagation conditions mapped against an infinite number of line-of-sight and nonline-of-sight paths with ground reflection and scatter and surface absorption components that need to be factored in to the model. Then the calculation must be repeated for K-band and Ka-band feeder links and then repeated again to assess interference levels with in band and adjacent band terrestrial 5G.

7.13 The Answer: Mixed Constellations Including 5G

The only plausible answer to this modeling problem is to remove the incentive to argue and this can only be done if the same entity owns and manages a GSO constellation and MEO constellation and LEO constellation and preferably 5G terrestrial assets as well. (Satellite TV broadcasting could be added to the mix as well.)

This would require seismic changes to existing competition policy and regulatory policy but would provide a uniquely powerful user experience and uniquely powerful platform for global IoT connectivity.

However, it is worth reflecting on the user experience benefits and economic gain that this approach would deliver. Let us imagine a mixed constellation as a combination of a high-count LEO, a medium to high-count MEO, and at least 4 GSO satellites. The four GSO satellites provide complete east to west coverage but could be scaled to theoretically 180 satellites assuming a presence at each available GSO orbital slot. The more orbital slots the better in terms of being able to point more or less directly upwards at the equator (rather than at a GSO on the western or eastern horizon). The LEO and MEO satellites provide north to south directly overhead coverage. The mixed constellation intersatellite switches both within each constellation but also between constellations. This reduces the number of ground Earth stations needed for feeder links. Note that the ground Earth stations are implemented as very large active flat antennas, which can service any satellite from horizon to horizon, although the most effective routing is always going to be directly upwards.

This also applies to all direct to Earth and direct Earth to space links supporting users and IoT devices where preference is always given to links that go directly up and down.

This means that Earth stations in motion (ESIMS) and stationary Earth stations can be equipped with low-cost passive flat VSATs that look upwards through a narrow cone of visibility, rejecting all unwanted energy at lower elevation angles.

At the equator, a passive flat VSAT looking upwards would see a GSO directly overhead all the time and a procession of LEO and MEO satellites passing overhead. A protocol would need to be established to manage how traffic would be multiplexed across these multiple satellites, but this is no different in principle to delivering traffic from multiple terrestrial base stations. Users and devices at higher latitudes have visibility either to a directly overhead high-count LEO or MEO.

As set out in Chapter 6, the combination of metamaterials and electronic band gap substrates means that antennas can be constructed that could potentially have efficient resonance from VHF to E-band. This would mean that the constellations could scale from 1-MHz by 1-MHz channel bandwidths and passbands at VHF to 5-MHz channel bandwidths in a 5-MHz passband at UHF (5G Band 31) through 10-MHz and 20-MHz channel bandwidths implemented in wider passbands up to 3.8 GHz (including the 3.4-GHz to 3.8-GHz 5G passband), then up through the extended C-band to Ku-band, K-band, and Ka-band (250-MHz channel bandwidths within 3.5-GHz passbands at 12 GHz, 18 GHz, and 28 GHz), then to V-band (40 and 50 GHz), and then to E-band (1-GHz channels in the 5-GHz passband either side of automotive radar at 77 GHz).

Note that this assumes that the LEO, MEO, and GSO constellations coshare the 5G refarming bands from 450 MHz to 3.8 GHz and that 5G and 5G in-band backhaul coshares the satellite core bands including extended C-band, the K-bands, V-band, and E-band.

The satellite sizes could scale from a few kilograms (CubeSats) to 60,000 kg (the largest single payload supportable on next generation SpaceX rockets to LEO). The RF power per satellite could scale from a few milliwatts (a LEO CubeSat) to a 50-kW GSO. The constellation count could scale from 180 GSO orbital slots to potentially hundreds of thousands of CubeSats.

We reference in Chapter 9 the Myriota CubeSats [9] that service the Australian outback using sub-1-GHz licensed spectrum with an uplink power budget of 33 mW with a flyover every 90 minutes but the subgigahertz industrial, scientific, and medical (ISM) bands could also be potentially used.

This combination of satellite and terrestrial spectral assets would deliver several desirable outcomes. There would be sufficient flux density and capacity to support smartphones and wearable devices from space. We take wrist-worn

GPS for granted, so this is a modest extra step (albeit one requiring an uplink) but one that helps terrestrial mobile operators deliver genuinely global coverage. Similarly, it provides satellite operators with the ability to become part of the consumer mass market value chain. The positive impact on mobile operator and satellite operator EBITDA and enterprise value would be substantial.

7.14 Up Before Down Constellations: Hubble Telescope and International Space Station as Prior Examples

As stated above, in an up before down constellation, a LEO or MEO uplinks to a GSO, which then downlinks the traffic to a GSO base station, with the reverse link following the same route. This adds link latency but reduces the number of Earth stations needed to support the MEO and LEO constellations. This is effectively an extension of intersatellite switching and provides greater flexibility in the way that traffic is handled.

The principle is not new. Communications with the Hubble Telescope (launched in 1993 into a LEO at 600 km) and International Space Station (launched in 1998 into a LEO at 400 km) go up to a GSO (the GSO Near-Earth Network) before coming back to Earth. This is known as the Near-Earth Network. The Near-Earth Network is part of the NASA Tracking Data And Relay Service, which was started in the early 1970s.

7.15 TRDS Protection Ratios

Incidentally, NEWLEOs such as OneWeb will need to coordinate with NASA to guarantee protection of TDRSS (Tracking and Data Relay Satellite Systems) in the 14.0-GHz to 14.2-GHz band and with space observatories operating radio astronomy services in the 10.6-GHz to 10.7-GHz band and with U.S. government satellite networks, both GSO and NGSO in Ka-band. These agencies are not generally motivated by the prospect of having to manage new sources of interference and have nothing to gain from saying yes.

7.16 Ground-Based Antenna Innovation (Passive and Active Flat VSATs) as the Enabler

It is worth restating at this point the importance of ground-based antenna innovation as the enabler of 5G and LEO, MEO, and GSO spectrum cosharing. Passive and active flat VSATs remove the need for progressive pitch, handover, and power control. Life becomes simpler in space and all those pesky arguments over interference go away. The alternative vertical straight up straight down option using antennas with a narrow, fixed cone of visibility directly upwards to directly overhead LEO, MEO, or GSO satellites or active beam steerable arrays, which can scan from horizon to horizon to provide best connect LEO, MEO, and GSO access provides all the mechanics needed for power-efficient spectrally efficient global coverage. These antennas can be integrated into LCD TV displays or solar panels to reduce cost and allow crossover functionality. For example, the TV screen in your living room can also function as an LTE and 5G array antenna, and the solar panels on your house roof and car roof can collect photons and electrons and support RF uplinks to directly overhead LEO, MEO, and GSO satellites.

This is easier to implement commercially if one entity owns or has access to all three delivery options (or five delivery options if you include 5G and 5G in-band backhaul). This approach does depend on the moderately contentious notion of the satellite industry cosharing spectrum with 5G terrestrial and in band backhaul at 12-GHz, 18-GHz, and 28-GHz V-band and E-band and the mobile broadband industry cosharing spectrum with the satellite industry between 450 MHz and 3.8 GHz and both industries coming to an amicable agreement on cosharing extended C-band. The mutual prize would be the potential availability of 16 and 32-element active arrays embedded in smartphone displays and wearable devices and higher-count element arrays (256, 512, 1,024) in larger form factor devices for very high throughput fixed and mobile applications and a completely different but ultimately sustainable 5G business model.

7.17 GSO HTS and VHTS Constellation Innovation

Meanwhile, GSO constellation innovation marches on. GSO constellation innovation is driven by multiple factors. As launch technology improves, larger heavier satellites can be lifted into GSO, which means that the amount of available RF power increases. For example, typical BIGSATs (very big satellites) today are somewhere around 6,000 kg with rockets typically able to lift payloads of the order of 10,000 kg. As stated above, next generation rockets are capable of delivering more than 60,000 kg into space. The satellites have ever more complex spot beam antennas that can focus RF power on discrete geographic areas and individual users and IoT devices within those areas and dynamically respond to changes in demand. The spot beams are communicating with Earth-based ground stations, which can be fixed or mobile with antennas that can actively discriminate wanted energy from unwanted energy. This increases throughput and improves the power efficiency and spectral efficiency and cost efficiency of the overall system. In the longer term, dish antennas will be replaced with flat panel active arrays providing more highly resolved terrestrial footprints, particularly at higher (V-band and E-band) frequencies.

Over the past 10 years, technical advances have enabled the introduction of the new generation of high throughput satellites (>1 Gbps of throughput) and over the next 10 years can be expected to enable the introduction of the next generation of very high throughput satellites in V-band and W-band (>10 Gbps of throughput). Fiber, cable, and copper and terrestrial cellular throughput are also increasing in parallel. In this next section, we set out to see how these advances might change how GSO constellations interact technically and commercially with the rest of the telecommunications industry.

First, we need to separate out the different commercial entities participating in the GSO market, specifically differentiating global satellite providers from regional providers and the sovereign SATs.

7.18 The Global GSOs

The global GSOs are entities that own and operate or have access to anything between 4 and 40 satellites in geostationary orbit (Figure 7.5). A minimum of four satellites will provide global coverage though higher-count constellations will provide more optimum coverage, with higher elevations west to east.

Note that while Inmarsat was originally focused on maritime coverage (hence the name) it now vigorously addresses terrestrial markets and has provided terrestrial services for over 30 years. Companies such as Intelsat that have traditionally focused on terrestrial markets are addressing maritime markets.

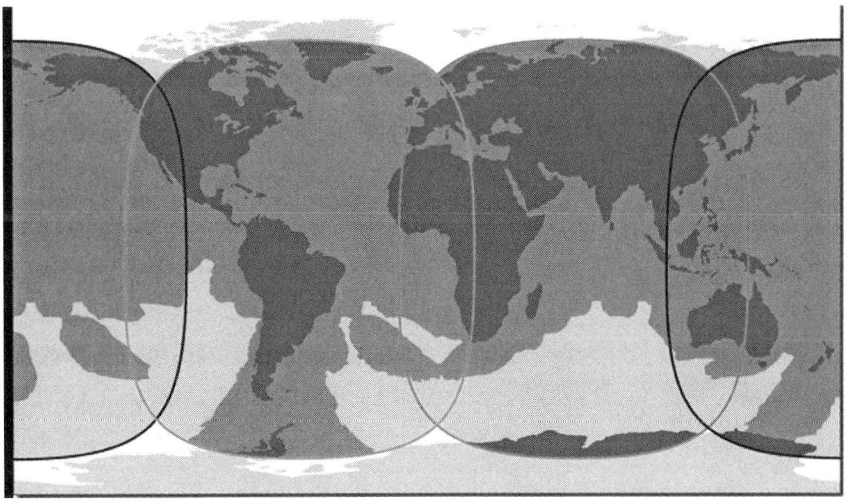

Figure 7.5 Global coverage from four GSO satellites. (Image courtesy of Inmarsat.)

The pictures from the Moon landing in 1969 were sent around the world via Intelsat GSO satellites. Intelsat was privatized in 2001, 2 years after Inmarsat. This coincided with the dotcom bust, the telecoms meltdown (Nortel going bankrupt), and the global fiber glut. In 2006, Intelsat acquired the Pan Am SAT becoming the world's largest fixed service satellite service provider. It has constellation assets in C-band, Ku-band, and Ka-band.

Intelsat provides an alternative to Inmarsat from a Ku-band constellation and launched the first of a new generation of six satellites in January 2016 via Europe's Ariane 5 heavy lift rocket. The new satellites weigh over 6,500 kg and are positioned to cover the Caribbean and North Atlantic cruise routes (and land masses as well). The satellites are manufactured by Boeing.

7.19 Other Global GSOs

Other global GSOs include Eutelsat and Viasat. Eutelsat, also now known as Eutelsat Ka SAT, was formed in 1977 to operate the first European satellites (launched in 1983). After the fall of the Berlin Wall, Eutelsat was extended to cover Eastern Europe. It was privatized in July 2001 and went through an IPO in 2005. TV still represents a substantive part of Eutelsat's income, broadcasting to the Middle East, Turkey, and Africa from the Hot Bird satellites. In January 2014, Eutelsat acquired SATMEX (Mexico), which effectively means that the company provides coverage of Europe, Africa, the Middle East, Asia, and the Americas. The EUTELSAT 7C electric satellite is scheduled to launch in the third quarter of 2018. The satellite will have 44 Ku-band transponders and will be cosited at 7° east position for Turkey and Sub-Saharan Africa, doubling the number of transponders from 22 to 42 over Sub-Saharan Africa. This means that Eutelsat will be offering capacity from more than 40 satellites. This represents a big share of the available geostationary orbit slots (even taking into account cositing) [10]. Viasat has expanded its footprint to deliver global coverage.

7.20 The Regional SATs

IP Star was the first of the regional high throughput satellites (launched in 2005). Other regional SATs include Arabsat, Hispasat, and Hylas (Avanti).

7.21 The Sovereign SATs

These include companies such as Telesat and national network constellations in Australia (part of Australia's National Broadband Network) and Singapore, Brazil, India, Indonesia, China, Iran, and Israel.

7.22 Very High Throughput Constellations

Very high throughput constellations are being proposed as the basis for specific new spectrum allocations at WRC 2019 at the 32-GHz band (31.8–33.4 GHz) and in the Q-band and V-band (37–52 GHz). This is described in WRC-19 agenda item 9.1.9.

SpaceX, OneWeb, Telesat, O3b Networks, and Theia Holdings all have submissions with the FCC for V-band satellites in nongeosynchronous orbits to provide communications services in the United States and elsewhere.

The Boeing submission is to use 37.5–42.5-GHz range of V-band for downlinking from spacecraft to terminals on Earth and 47.2–50.2 GHz and 50.4–52.4 GHz for linking back to the satellites using 1,396 to 2,956 LEO satellites.

SpaceX proposes a VLEO, or V-band LEO constellation of 7,518 satellites to follow the initial proposal for 4,425 satellites that would function in Ka-band and Ku-band. Canada-based Telesat described its V-band LEO constellation as one that "will follow closely the design of the Ka-band LEO Constellation," using 117 satellites not counting spares as a second-generation overlay.

Theia asked the FCC to allow it to use V-band frequencies for gateways on the ground that would have originally only used Ka-band.

OneWeb wants to operate a subconstellation of 720 LEO V-band satellites at 1,200 km, and another constellation in MEO of 1,280 satellites expanding the OneWeb constellation by 2,000 satellites. Traffic would be dynamically assigned between the LEO and MEO V-band constellations based on service requirements and the data traffic within coverage areas.

OneWeb's application for MEO follows that of Viasat in November for 24 MEO satellites to augment Viasat 3, the company's trio of terabit-per-second-throughput satellites currently under way. Viasat bundled its request for use of V-band together with its application for MEO Ka-band. O3b wants market access to V-band for up to 24 additional satellites that would operate in a circular equatorial orbit as a constellation called O3bN.

7.23 Autonomous CubeSats

We earlier covered CubeSats, but is is worth briefly revisiting some of the more active particpants in the sector. Whereas GSO satellites tend to expand coverage by longitude, CubeSats expand their coverage by latitude. Figure 7.6 shows the initial equatorial coverage from the proposed SAS CubeSat constellation (200 satellites each weighing 10 kg).

The proposed SAS constellation has onboard orbit control and autonomous network management, which makes the space segment independent from terrestrial control. The constellation proposes to use four-way intersatellite links

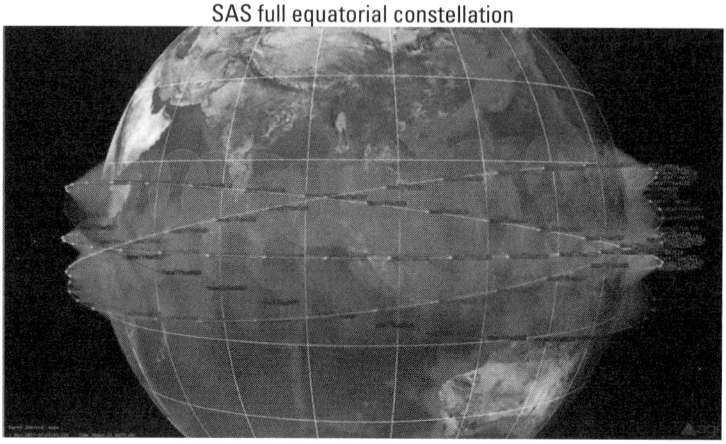

Figure 7.6 Sky and Space Autonomous Satellite Constellation. (Thanks to SAS.)

(up, down, and side to side) with the links providing real-time 3-D satellite to satellite distance estimation providing the basis for automated autonomous station keeping. Figure 7.7 illustrates the size and shape of the satellites and the way in which they can be stacked on top of each other.

7.24 Space-Sensing Constellations: Square Satellites That Look Around

Our focus in this book is on communication satellites, but there is a parallel increase in the use of satellites for imaging and sensing [11]. The Planet Labs Rapid Eye constellation [12] is a present example. Five 1-m^3 satellites each weighing 150 kg in a 630-km Sun-synchronous orbit collect spectral information from 440 nm (blue) through to near infrared (760–850 nm) cross an area of 6 million m^2 per day. The constellation is used to monitor illegal deforestation and for agricultural and energy and infrastructure monitoring.

7.25 GNSS Satellites

As with space sensing, we are not attempting to cover positioning and location satellites in detail except to state the obvious. GPS satellites fly in MEO at an altitude of 20,200 km. Each satellite circles the Earth twice a day [13].

There is also Galileo, the European version of GPS and Beidou, the Chinese version of GPS, and IRNSS, the Indian Regional Navigation Satellite System.

Figure 7.7 Sky and Space Global CubeSat.

Positioning and location are also available from other MEO constellations (O3b) and LEO constellations courtesy of their strong Doppler signature and (relative to GPS) high flux density. The Iridium Next constellation service offer includes positioning, timing and authentication to augment GPS technology for critical applications.

7.26 Quazi Zenith Constellations

Augmented GPS is also available from the Japanese Quazi Zenith constellation (quazi zenith is the term used to describe geosynchronous as opposed to geostationary orbits), initially from three geosynchronous satellites and one geostationary satellite (Mitchibiki) all broadcasting same L1to L6 signals as GPS and other GNSS systems [14].

Note that being geosynchronous rather than geostationary means that the satellites progress in a figure-of-eight pattern over Japan.

7.27 Orbital Debris

More than 500,000 pieces of debris or space junk are tracked by radar as they orbit the Earth. This debris can travel at speeds up to 17,500 mph, fast enough for a relatively small piece of orbital debris to damage a satellite or a spacecraft. The rising population of space debris increases the potential danger to all space vehicles. Such debris includes nonfunctional spacecraft (Sputnik I is still up there), abandoned launch vehicle stages, mission-related debris, and fragmentation debris. Old satellites can contribute substantial debris fields if any hydrazine propellant is not vented at the end of life; eventually (through collision with debris or by repeated thermal cycling), the fuel tanks fracture explosively.

There are many millions of pieces of debris that are so small that they cannot be tracked. Even tiny paint flecks can damage a spacecraft when traveling at these velocities (space shuttle windows have been replaced because of damage caused by paint fleck impacts).

There is relatively little debris in GEO and MEO orbits, but substantially more in LEO orbits. Two major events substantially increased the amount of LEO orbital debris.

On February 10, 2009, a defunct Russian satellite collided with and destroyed a functioning U.S. Iridium commercial satellite. The collision added more than 2,000 pieces of trackable debris to the inventory of space junk.

China's 2007 anti-satellite test, which used a missile to destroy an old weather satellite, added more than 3,000 pieces to the debris problem.

The position of the International Space Station is regularly adjusted to avoid potential collisions with larger items.

Geostationary satellites are spaced just over 75 km apart and debris-related failures are rare (once every decade). Older spacecraft are parked at a higher orbit to avoid potential collisions (and to allow their orbital slot to be reused).

The risk of more collisions in LEOs will clearly increase with the increase in LEO satellite deployments, and debris management will become a significant potential cost overhead for the NEWLEO operator community.

Guidelines have been proposed to ensure that satellites at the end of their lives can be vented (passivated) and either deorbited in a controlled fashion (ideally for LEO satellites to burn up in the atmosphere, although this is not necessarily possible with larger satellites) or placed in safe orbits. The majority of nations launching satellites (and the United Kingdom) have now signed up to these; however, China is not a signatory.

Initiatives to actively reduce the amount of space debris have been proposed by a number of organizations, and the European Space Agency (ESA) is active in this field. This is rather similar to the problem of dumping plastic at sea over several decades. It is a cumulative problem with a solution that requires everyone who has contributed to the mess to help clear it up, which is unlikely to happen.

7.28 Subspace High Altitude Platforms

High-altitude platforms fly around above terrestrial networks and below space networks and potentially share the same spectrum; they are definitely part of the overall system to the system interference that we have been describing.

The sector divides into heavier than air (HTA), which includes the Aquila drone developed by Facebook [15]. The drone flies at between 60,000 and 90,000 feet (above the weather).

The plane has a wingspan rather wide than a Boeing 737 and is intended to provide Wi-Fi connectivity across a 60-mile diameter/30-mile radius cell. It runs on solar power and batteries and needs about 5 kW to keep it up in the sky providing downwards RF power. It is planned to remain air born for 90 days at a time or longer if possible as landing and taking off is hard work and perilous for what is essentially a fragile piece of machinery. It travels at a stately 80 miles per hour mainly in circles.

7.29 Lighter-Than-Air Platforms

Lighter-than-air platforms include updated versions of the R101 [16] and Zeppelin [17]. This includes products such as the Airlander, financed by the lead singer from Iron Maiden, Bruce Dickinson [18].

Lockheed has a similar project [19], although without heavy metal funding.

These are intended as multipurpose platforms carrying people and materials to remote places quite slowly. They fit in to the category of the Internet of Large Slow Moving Objects (IoLSMO) so could plausibly become part of an overall integrated communications platform although weather can be quite a problem.

Arguably, the highest profile experiments have been with balloons under the Google Loon project. Like the Aquila drones, they are designed to fly above the weather at around 60,000 feet.

High-altitude platform systems could potentially be an efficient way of providing temporary coverage or additional capacity. A high-altitude platform system quasi-stationary stabilized platform at an altitude of between 8 and 50 km could serve a 200-km diameter cell though the elevation angle at the cell edge would be 10°, which could result in unacceptable blocking from terrestrial buildings and foliage.

7.30 Summary

Adding all the FCC filings together produces a grand total of over 10,000 NEWLEO satellites planned for launch into a bewildering selection of LEOs with the satellites scaling from a few kilograms to 1,000 kg. The overall aim is to replicate the high throughput service offers from incumbent GSO operators in Ku-band, K-band, and Ka-band and provide a path to very high throughput service offers in V-band and W-band. High-latitude coverage will also be better than GSO.

This performance leap is to be achieved by using the same spectrum as the GSO operators, typically 3.5 GHz + 3.5 GHz passbands in Ka-band or potentially 5 GHz + 5 GHz passbands in V-band and W-band (E-band) combined with a high-count constellation, with each constellation supporting hundreds or potentially thousands of satellites.

This seems like a wondrous prospect though even thousands of satellites is a numerically small number compared to the millions of installed LTE base stations (Huawei claimed to ship over 1 million LTE base stations per year).

Interestingly, this is all technically possible and at least some of the constellations could be financially viable. The main barrier to deployment is not technical but regulatory and revolves around the knotty subject of interference mitigation and protection ratios.

There are technical solutions to interference management, but they have an associated cost. High GSO protection ratios will mean that NEWLEOs will need to reduce output power and deliver downlinks at nonoptimum elevation angles with long path lengths through the atmosphere. This will reduce throughput and capacity and introduce additional latency all of which will reduce the viability of the NEWLEO constellation offer.

Ultimately, this can probably only be resolved by creating corporate entities who either own or have access to all of the constellation options including LEO, MEO, GSO, and 5G terrestrial integrated with 2.4-GHz and 5-GHz Wi-Fi, but this requires a substantive shift in competition policy across all addressable geographic and vertical markets.

More radically, a new generation of antenna technologies based on metamaterials and electronic bandgap substrates produces a new class of antennas that we describe as flat VSATs (although they can also be conformal). Flat VSATs can be realized as active and passive arrays. Passive arrays look directly upwards through a narrow cone of visibility. Active arrays scan from horizon to horizon to provide the best connect access. The antennas can have wavelength resonance from VHF to E-band.

Flat VSATs remove the need for progressive pitch and power control (handover or satellite selection may still be needed at the equator), but essentially flat VSATs remove the tension points between NEWLEO and legacy MEO and GSO operators by resolving space and ground Earth station interference issues.

They also provide the opportunity for the 5G community to coshare 5G sub-3.8-GHz spectrum with satellite operators and enable satellite operators to coshare K-band, V-band and E-band spectrum (12, 18, 28, 40, 50 GHz and the 5-GHz duplex either side of the 77-GHz automotive radar band) with 5G.

This would improve spectral and power efficiency but also transform the delivery economics of both industries. The emergence of subspace options such as the Facebook drone project and Google Loon project demonstrate that the GAFA quartet (Google, Apple, Facebook, and Amazon) has an appetite for investing in nontraditional connectivity platforms. This is coupled with their developing appetite for launch technology investment (see Chapter 6).

The GAFA quartet and other Web-scale companies have two other significant advantages: cash and customers. This introduces additional scale into the constellation investment equation, the focus for our next two chapters.

References

[1] http://www.unoosa.org/pdf/limited/c2/AC105_C2_2014_CRP15E.pdf.

[2] http://spacenews.com/three-orbcomm-og2-satellites-malfunctioning-fate-to-be-determined/.

[3] https://www.orbcomm.com/en/company-investors/news/2017/orbcomm-acquires-inthinc.

[4] https://www.fcc.gov/document/fcc-grants-oneweb-us-access-broadband-satellite-constellation.

[5] https://www.o3bnetworks.com/wp-content/uploads/2015/02/O3b-Technology-Overview-A4_10SEP14.pdf.

[6] http://spacenews.com/oneweb-gets-slide-decked-by-competitor-at-casbaa/.

[7] http://www.absatellite.com/company/corporate-overview/.

[8] http://www.casbaa.com/.

[9] http://myriota.com/.

[10] http://www.eutelsat.com/en/services/data/consumer-broadband/tooway.html.

[11] https://ineke.co.uk/2013/01/20/charlie-and-the-chocolate-factory-by-roald-dahl/.

[12] https://www.planet.com/products/satellite-imagery/rapid-eye-basic-product/.

[13] www.gps.gov/systems/gps/space/.

[14] http://qzss.go.jp/en/.

[15] https://www.facebook.com/notes/mark-zuckerberg/the-technology-behind-aquila/10153916136506634/.

[16] http://www.airshipsonline.com/airships/r101/Crash/R101_Crash.htm.

[17] http://www.airships.net/zeppelins/.

[18] https://www.hybridairvehicles.com/news-and-media/news/business-insider.

[19] https://www.lockheedmartin.com/us/products/HybridAirship.html.

8

Production and Manufacturing Innovation

8.1 Aviation Manufacturing: A Fairy Tale

In a previous incarnation, I sold spot welding equipment. The Hawker Siddeley Company in Kingston was a customer. The company employed 5,000 skilled craftsmen to assemble parts for the Harrier Jet including a small army of men in leather aprons hand-fettling titanium cowlings with small hammers.

In 1963, Prime Minister Harold Wilson gave a famous speech about "The White Heat of Technology," talking about how Britain would remain at the forefront of technology and manufacturing innovation. The Hawker Siddeley Factory (Figure 8.1) could best be described as the Snow White Heat of Technology.

Forty years later, the design and manufacturing of fighter jets remains based on materials and manufacturing innovation. Figure 8.2 shows the BAE Titanium manufacturing facility in Australia.

The companies that build fighter jets also build satellites. Hawker Siddeley, for example, was nationalized in 1977 and became British Aerospace and in 1999 became BAE systems [1]. BAE is a joint contractor with Lockheed Martin on the F35 fighter jet, a $1.3 trillion project.

Figure 8.1 The Hawker Siddeley Head Office in Kingston, now a luxury housing estate.

Figure 8.2 BAE Titanium production facility.

8.2 Satellite Manufacturing: A Similar Story?

Is satellite manufacturing still a craft industry? To an extent, yes. The satellite industry has spent the last 60 years servicing customers who either cannot be connected any other way or who find it difficult to connect via terrestrial networks. The industry is currently based on a low-volume (both for satellites and terminals)/high-cost business model and has been able to be keep device and access prices at a relatively high level when compared to terrestrial networks.

The satellite industry supply chain has always been sustained by high added-value defense work, and this remains a dominant source of margin and profit. It also means that the supply chain is not geared up to deliver low-cost devices or low-cost network hardware and software and system support. Companies like Lockheed Martin or Boeing or Airbus or Thales or Hughes or

Northrop Grumman are not structured to deliver equipment at consumer price points.

However, this started changing 20 years ago when Motorola set out to source the first generation of Iridium satellites using what at the time was a large number of satellites (66 satellites + 6 orbital spares) as the basis for a horizontally integrated moving production line, which by 1997 was capable of producing a satellite in 4.3 days. Motorola had the advantage of significant experience in handset and base station manufacturing including a rigorous Six Sigma approach to quality, which was transferred very effectively across the production line. Note that Six Sigma was explicitly a quality standard aimed at reducing build cost both for the satellites and back through the supply chain [2]. The fact that the satellites lasted three times as long as expected (21 rather than 7 years) suggests that this marked a significant move forward in satellite manufacturing. This was achieved without the added cost of a clean room environment or use of traditionally space qualified components. Boeing is using similar production-line techniques for replacement GPS satellites, albeit at lower volumes.

Twenty years later, Iridium has produced their NEXT generation satellites at the Orbital ATK Satellite Manufacturing Facility in Gilbert, Arizona, under the supervision of the lead contractor Thales with a total production run of 81 satellites (66 + 9 orbit spares + 5 ground spares). Orbital ATK was acquired by Northrop Grumman in October 2017.

Thales was also the lead contractor for the replacement satellites for the Globalstar constellation. Globalstar claims that the buildout cost for the replacement at $1 billion was 20% of the buildout cost of the original constellation ($20 billion). Part of this saving can be ascribed to lower production and manufacturing costs.

OneWeb stated that it needs to produce 15 satellites a week from the Toulouse and Florida factories [3], a manufacturing joint venture between OneWeb and Airbus.

Figure 8.3 follows through the assembly stages of these two facilities. Each factory has 30 work stations with two staffs at each station. Collaborative robots (cobots) help with local lifting tasks. Smart tools are used that recognize the bolts on which they are placed and the tightening torque required, automatic optical inspection checks for any assembly misalignment.

The facilities will work one shift a day to fulfill the initial requirement of 648 satellites, although this could be increased to three shifts. This would provide sufficient capacity for OneWeb and Airbus to offer production facilities to third parties.

Generally, it can be stated that the satellite industry has a legacy supply chain with profits derived primarily from military work. This is not a supply chain optimized to produce devices and networks at commercial price points.

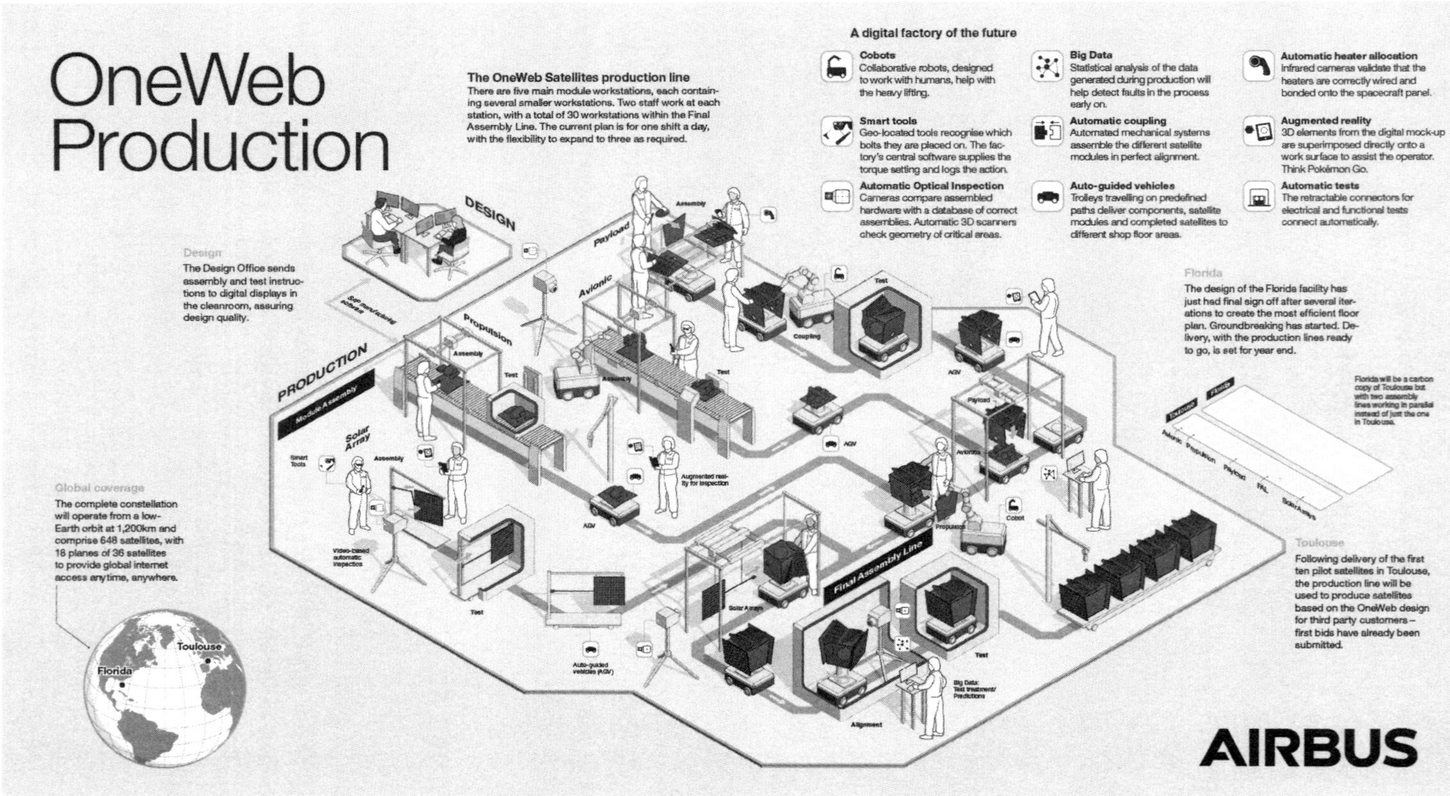

Figure 8.3 OneWeb Toulouse and Florida manufacturing facilities. (With thanks to Airbus.)

The OneWeb/Airbus initiative aims to change that. The finished product is shown in Figure 8.4.

8.3 The Automotive Industry as a Source of Satellite Manufacturing Innovation

8.3.1 Mr. Ford and Mr. Musk

In my previous book, I referenced Mr. Ford as the doyen of low-cost, high-quality volume manufacturing with the Ford Model T being the prime example of performance gain through close management of production tolerances combined with the use of vanadium steel (lighter and stronger) and a meat hook production line copied from the local abattoir.

One hundred and ten years later, Mr. Musk can almost certainly transfer the lessons learned manufacturing Tesla cars and batteries into his SpaceX constellation venture.

8.3.2 Production Innovation for 5G Smartphones: Why Scale Is Important for Performance

The equivalent automated production line in a 4G and 5G user device and base station factory would be a state-of-the-art pick-and-place machine. Figure 8.5 shows a Panasonic machine capable of placing 100,000 components per hour. To put this in perspective, the last time I walked around a (3G) production plant (San Diego, 2002), the fastest pick-and-place machine could manage

Figure 8.4 OneWeb satellite. (With thanks to Airbus.)

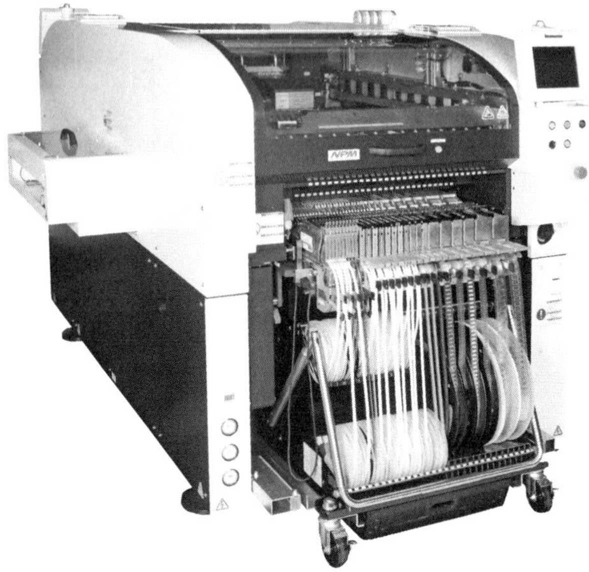

Figure 8.5 Panasonic pick-and-place machine.

7,000 components an hour. This machine can place anything from exceedingly small and narrow 0402 (1005) mm (inch) and 1005 (0402) mm (inch) chips to 90 × 100 mm chips on boards up to 18 inches by 20 inches [4]. It achieves this by using high-resolution fast imaging.

Note that, as with the automotive industry, scale is needed in order to invest in manufacturing technology to not compromise component performance to not compromise product performance.

This is particularly important for radio frequency (RF) active and passive components at higher frequencies and for high-performance static sensitive memory products. Taping systems are used to place components in the correct orientation and position and order and condition (protection from damp and static) on a tape that is then fed in to the pick-and-place machine.

Note that scale effectively confers performance gain. I referenced Nokia's ability 15 years ago to produce GSM phones within minutes, but Nokia also used its production scale to bully its supply chain into delivering more tightly toleranced RF components. The result was a steady year-on-year improvement in RF performance. A GSM phone in 1992 from any vendor barely based the conformance specification threshold of –102-dBm sensitivity. By 2002, Nokia phones could be measured with receive sensitivity of the order of –109 dBm.

Much of this manufacturing expertise has now moved to China, but remains critically important.

8.3.3 Materials and Manufacturing Innovation in the 5G Supply Chain

The same principle applies to the supply chain, particularly the RF component supply chain, particularly that for centimeter-band and millimeter-band smartphones and IoT devices.

In 1982, it was a major challenge for the industry to produce cellular phones at consumer price points that could work at 800 MHz (the United States) or 900 MHz (Europe). For example, standard FR4 printed circuit board material was barely adequate [5]. Over the following 35 years, the industry had to accommodate higher frequencies, initially 1,800/1,900 MHz, then 2 GHz, then 2.6 GHz, and then 3.4 GHz. As frequency increases, RF gain becomes more expensive and noise becomes more problematic. It also becomes harder to switch and filter RF signals. As always, a technical challenge became a commercial opportunity. Higher-frequency radio and radar systems were widely used in military radio and radar systems but used more exotic materials such as gallium arsenide for RF power amplifiers and low-loss linear switch paths. These materials required innovative manufacturing techniques. The additional cost of the materials compared, for example, to basic silicon meant that it was particularly important to maximize yield (the percentage of devices meeting the agreed performance specification).

Companies such as Rockwell Semiconductor [6] translated these material innovations and manufacturing techniques into new companies specializing in supplying RF components for 3G and 4G smartphones. Put in biblical terms, Rockwell Semiconductor begat Conexant [7] which begat Skyworks, which today is a major supplier of 4G RF power amplifiers. Hewlett Packard Semiconductors (founded in 1961) begat Agilent Technologies (1999), which begat Avago [8], which is a major supplier of acoustic filters and switches to the industry (and recently acquired the RF assets of Broadcom). RF Micro Devices and TriQuint became Qorvo [9], and Peregrine Semiconductor morphed into Murata (RF switch products and filters for 4G smartphones).

It is worth noting that all these companies are U.S. companies. They also supply the military radio and radar market and are an increasingly important part of the automotive industry supply chain. These companies regard 5G and particularly 5G products implemented in the centimeter and millimeter bands as a critical target market.

8.3.4 Materials and Manufacturing Innovation in the Rocket Industry

In Chapter 4, I referenced Rocket Lab USA as an example of materials and manufacturing innovation with the use of carbon fiber composites for the rocket shell and three-dimensional (3-D) printing (additive manufacturing) for the rocket motor.

Rockets are characterized by the need to produce large containment tanks that can be filled with very explosive liquids at high pressure without the risk of leaking or structural failure.

This used to be very hard to achieve using traditional welding techniques. In the SpaceX rockets, for example, the first-stage propellant tank walls are made of aluminum lithium and are manufactured using friction stir welding [10], a technique invented in 1991 by TWI Limited where a weld is made as the name implies by a process of friction and rotation with a rotating tool head positioned between two clamped plates. The frictional heat causes a plasticized zone to form around the tool. The rotating tool moves along the joint line forming a consolidated solid-phase joint, avoiding the shrinkage, solidification, cracking, and porosity associated with fusion welding (see Figure 8.6).

8.3.5 Meanwhile, Back at the Battery Farm

Mr. Musk meanwhile has also turned his attention to battery manufacturing (coincidentally also with Panasonic). To complete the picture, Mr. Musk also has an impressive solar panel factory in Buffalo, New York.

The investment by Mr. Musk in automotive manufacturing, solar panel manufacturing, and battery manufacturing provides him with many of the manufacturing skill sets needed to manufacture satellites. This is presumably potentially complementary to the SpaceX investment in rocket launch technology and constellation technology.

Figure 8.6 Friction stir solid state welding. (With thanks to TWI Limited [formerly The Welding Institute].)

8.3.6 Automotive Enterprise Value: Mr. Musk as a Modern Marconi

As a matter of record, Ford produces 6.6 million cars a year and makes a profit on every car. Ford has an enterprise value of $46 billion. Tesla produces 120,000 cars a year and makes a loss on every car. Tesla has an enterprise value of $48 billion. This enterprise value is used as a mechanism for raising capital, which is used to invest in manufacturing technology and manufacturing assets (factories). Conveniently, these factories are built in areas that are politically important, Florida, for example.

In many ways, Mr. Musk is the modern Marconi, the man who "networked the world," the supremely effective self-publicist adroit at turning disaster (the sinking of the *Titanic* in 1912) to commercial advantage.

8.4 Automotive Radar Supply Chain as a Source of Satellite and 5G Antenna Manufacturing Innovation

The automotive industry also has visibility to the automotive radar supply chain. Companies such as Delphi Technologies have been producing automotive radars now for 20 years [11].

Present-day products include short-range, mid-range, and long-range radar products with very similar angular power detection requirements to 5G AES (Adaptive Electronically Steerable) antenna arrays and similar algorithmic processing requirements for angle of arrival, speed, and distance of nearby slow and fast-moving objects (and stationary objects as well).

In common with other automotive safety product vendors, Delphi supplies a range of radar based. Many LIDAR and imaging-based products have potential applications, particularly in space, for example, for docking systems and for self-autonomous satellites.

8.5 Supply Chain Comparisons

We have said that traditionally companies manufacturing satellites have income predominantly derived from military markets. These companies have research and development budgets paid for almost entirely by the defense community. By contrast, the mobile operator vendors, Huawei, Ericsson, and Nokia, have research and development budgets of 12% to 14%, similar in actual and ratio terms to the major automotive companies (Table 8.1).

It can be seen that the satellite industry is two orders of magnitude smaller in turnover. In terms of customer reach, there are over 4 billion cellular phone users. The number of users of satellite products depends on who and what you count. If you include satellite TV, it is a big number, although nowhere near 4

Table 8.1
Supply Chain Comparisons

5G and Satellite Industry Supply Chain Scale Comparisons–Some Examples										
5G Mobile Operators				**5G Network Vendors**		**5G Device Vendors**		**5G Component Vendors**		
	AT&T	Verizon	Vodafone	Ericsson	Nokia	Samsung	Qualcomm	Broadcom	Skyworks	Qorvo
Turnover	$150 billion	$131 billion	$60 billion	$23 billion	$16 billion	$160 billion	$20 billion	$15 billion	$3 billion	$2.4 billion
Employees	250,000	183, 000	111, 000	116, 000	140, 000	275, 000	27, 000	18, 000	6,000	7,000
Satellite Industry—the SATs						**Satellite System Vendors**				
	Inmarsat	Intelsat	Eutelsat	Arabsat	Echostar	Boeing	Lockheed	Northrop	Raytheon	Harris
Turnover	$1.3 billion	$2.35 billion	$1.4 billion	$1.5 billion	$3 billion	$96 billion	$46.8 billion	$24 billion	$23 billion	$7.5 billion
Employees	1600	1150	1000	2000	3000	160,000	98,000	65,000	61,000	21,000
Other Stakeholders—The GAFA Quartet (aka Over the Top/OTT companies and/or Webscale Internet Companies) and Microsoft										
	Google	Apple	Facebook	Amazon		Microsoft				
Turnover	$75 billion	$240 billion	$18 billion	$107 billion		$100 billion				
Employees	57,000	115,000	13,000	230,000		118,000				
Other Stakeholders—The Automotive Industry										
	Toyota	VW	GM	Ford	Tesla					
Turnover	$260 billion	$245 billion	$160 billion	$140 billion	$7 billion					
Employees	330,000	610,000	200,000	199,000	14,000					
Test Equipment Vendors—Servicing All of the Above Sectors										
	Keysight	R and S	Anritsu							
Turnover	$2.9 billion	$1.83 billion	$0.8 billion							
Employees	9500	9900	4000							

billion. If you exclude satellite TV, served users and devices in the satellite sector are counted in the order of millions or, at most, the low tens of millions.

As profiled in Chapter 7, the GAFA quartet has two major assets, cash and customers. Google has over 1 billion users, Facebook has 2 billion users, Amazon has 65 million Prime users, Apple has 588 million users who own 1 billion devices (1.7 devices per user), and PayPal (founded by Mr. Musk) has over 200 million registered accounts [12].

8.6 Why Scale Is Important

These numbers are important because of the impact of scale on user and IoT device availability, functionality, and cost. Let us take a timeline of 40 years and go back to 1977, coincidentally the last time I visited the Hawker Siddeley factory. This was the before cellular era with minimal market volume. Low-cost cordless phones were hand-assembled and hand-soldered. In 1987, life had moved on but it still took Motorola 8 hours to manufacture and test a cell phone.

The big shift came between 1992 and 2002 as GSM volume increased. By 2002, Nokia was manufacturing GSM phone in minutes and owned the components in the phone for less than a minute, an exemplary example of supply chain optimization and control.

In 2007, Apple introduced the iPhone. The success of this product over the next 10 years has been extensively documented but can be summarized as the driver behind Apple's ascendance to a dominant position in the mobile broadband value chain, only challenged in the user device market by Samsung.

Notably, Google has found it far harder to emulate this success story. In 2012, it acquired Motorola Mobility for $12.5 billion in cash. This was subsequently sold to Lenovo for $3 billion 2 years later. In 2017, Google paid just over $1 billion to acquire the division of Taiwan's HTC Corp that had developed the U.S. firm's Pixel smartphones, but the success of this venture has yet to be proven. HTC's worldwide smartphone market share declined to 0.9% in 2017 from a peak of 8.8% in 2011 and Google's Pixel had less than 1% market share since it was launched a year ago, with an estimated 2.8 million shipments.

This highlights an important point for the satellite industry. There are confident assertions within the industry that the electronic subscriber identity module (eSIM) [13] will change everything and all that is needed to enable satellite connectivity in next generation smartphones is a downloadable app.

In 2011, Apple was granted a patent for an eSIM that could be used to create a virtual mobile network (VMN). An eSIM replaces a physical SIM card (a piece of plastic with a bit of memory and microcontroller on it) with a server-based virtual SIM. The eSIM has been technically feasible for 20 years but resisted by the mobile operator community on the basis that it would make it easier for third parties to own and control customers, a legitimate concern.

However, there are two evolving product sectors where the hardware SIM becomes impractical. The first sector is the market for low-cost IoT connectivity where the initial cost of the SIM is a problem. Additionally, you cannot post a replacement SIM to a device and expect the device to install it.

The second sector is the emerging market for high-value consumable wearables. The Apple Watch Series 3 is the highest-profile contemporary example. Even the smallest SIM would take up too much space.

In addition to the eSIM, Apple Watch introduces several significant material and manufacturing innovations. LTE connectivity in the device is achieved by using the screen as an antenna; the device is water-resistant to 50m and includes a barometric altimeter, a GPS receiver, power-optimized Wi-Fi, and Bluetooth Low Energy (the battery needs to last all day for the device to be useful). The Apple Watch Series 3 is therefore a significant example of how technology innovation, combined with materials and manufacturing innovation, can be translated into additional user value [14]. While there are no announced plans to provide satellite two-way connectivity into this device, we are as users accustomed to having watches where all the functions work wherever we go. The challenge for the satellite industry is to realize this additional connectivity within the existing form factor of the device. This, in turn, will be a function of the RF bands supported and the physical layer compatibility with present and future 4G and 5G connectivity.

As discussed in Chapters 2 and 7, some of the satellite passbands, for example, in L-band and S-band, are adjacent to LTE spectrum but it will be hard to motivate smartphone manufacturers, particularly the two largest manufacturers by volume and value (Apple and Samsung) to extend existing passbands or add another switch path, both of which will add cost and compromise performance, particularly in products like the Apple Watch where space and energy consumption are at a premium.

This would only be justifiable if Apple or Samsung had a direct fiscal interest in adding satellite connectivity. There is no present evidence for either of these two companies that they would wish to even consider this as an opportunity, although this may change. However, the underlying narrative of this book is that it is possible to add satellite connectivity, specifically in band satellite connectivity from potentially ultrahigh frequency (UHF) (Band 31) to E-band (either side of the 77-GHz automotive radar band). The motivation is that devices such as the Apple Watch will work anywhere and everywhere on Earth including the land mass and oceans from east to west and north to south with sufficient spectral density available at all locations to deliver mobile and fixed broadband connectivity. Surely this must be useful added value?

8.7 Production and Manufacturing Challenges of Centimeter-Band and Millimeter-Band Smartphones

The alternative would be to look at hybrid satellite smartphone RF hardware in the centimeter and millimeter bands. This would depend on the satellite operators being willing to share the spectrum, which is an initial hurdle.

The benefit of these shorter wavelength bands is that antennas are more compact, but there are many manufacturing issues associated with high-volume RF products at these shorter wavelengths.

These can be summarized as:

- Loss due to the surface roughness of printed circuit boards: any surface roughness on printed circuit boards can result in significant losses.
- Parasitic capacitance effects: There are many parasitic capacitance effects that need to be managed both on the printed circuit board (PCB) and through all switch paths.
- Lower RF amplifier power-added efficiency: a GSM Class C power amplifier at 1 GHz has a power-added efficiency of 50%. This falls to 10% for a Class A amplifier at 28 GHz.
- Higher noise LNA: Low noise amplifiers are harder to realize in terms of their noise performance, gain, dynamic range, and power drain.
- Filter performance is harder to achieve: At frequencies over 3.8 GHz, acoustic filters become fragile and need to be replaced with ceramic filters. The good news is that ceramic filters get smaller as frequency increases (they are too big and expensive to use at lower frequencies). However, normal ceramic filters are fired at a temperature of at least 1,500°C and therefore need fire-resistant tungsten or molybdenum electrodes, which have relatively high electrical resistance. This causes unacceptable signal delay at higher frequencies. The answer is to use low-temperature cofired ceramic (LTCC) materials, which mix glass with alumina ceramic to reduce the firing temperature to less than 900°C. This allows copper or silver to be used to connect the internal layers of the device [15]. New generations of devices include inductors and capacitors in the filter package to reduce height and device real estate footprint.
- Modeling: Modeling tools are less well established
- Test and measurement: These can be harder to manage and connectors and cabling need to be more closely specified.

All the products available in these bands at present, for example, point-to-point backhaul products, are low-volume, hand-assembled, and individually optimized. This is an expensive hill to climb and you need specific climbing skills to get to the top both in terms of RF materials innovation and manufacturing innovation.

8.8 Wi-Fi, Bluetooth, or Subgigahertz IoT Connectivity as an Option

The alternative is to provide Wi-Fi connectivity, Bluetooth connectivity, and/or subgigahertz IoT connectivity. SAT-Fi products already exist, although none are as yet at price levels where mass market adoption in developing economies could be achieved without significant subsidy.

Note also that Wi-Fi is a low-power (10 mW) radio designed for local area connectivity and does not scale to larger-diameter cells partly due to the low transmit power, partly due to limited sensitivity (a function of the TDD physical layer) and partly due to TDD intersymbol interference in larger cells.

This is the reason why the Apple Watch 3 has an LTE transceiver. Admittedly, it also has a GPS receiver, but this is a low data rate receive only physical layer. Getting a satellite uplink to work technically and commercially in an Apple Watch or in similar products in this emerging wearable device sector is going to be a technical and commercial and regulatory challenge for the satellite industry but one that can be achieved from multiple constellations cosharing spectrum with 5G.

At the time of this writing, smartphones that support 802.11ad 60-GHz Wi-Fi were beginning to be introduced [16]. This represents a significant step forward in terms of design and producing higher-power, wider-area, millimeter-band phones at consumer price points.

8.9 Access Points and Base Station Hardware

Last but not least, it is important to recognize that the LTE base station business is now a high-volume business scaling from low-cost pico and micro base stations to surprisingly low-cost macro base stations where the site costs substantially outweigh the RF hardware and baseband hardware and antenna costs.

LTE base stations are manufactured on highly automated production lines in volumes that are in the tens and hundreds of millions per year. The ambition will be to leverage this scale into 5G base station and access products in the centimeter band and millimeter band. In-band backhauling would further consolidate this scale gain and help to amortize centimeter-band and millimeter-band RF research and development.

8.10 Server and Router Hardware Manufacturing Innovation

The three major LTE vendors, Huawei, Ericsson, and Nokia, are also heavily invested in next generation server and router hardware. The general assumption is that server hardware is largely commoditized with a limited number of processor and memory suppliers who closely control component production and manufacturing.

It is possible that new hardware architectures will emerge in the future. These could create new software optimization opportunities. Quantum computing is a possible candidate but remains curiously dependent on manufacturing innovation (resolution of the noise problem).

In our next chapter, we also discuss edge computing where server bandwidth is moved to the edge of the network. The motivation is partly to deliver the millisecond latency requirements embedded in the 3GPP standards and to reduce backhaul loading and traffic through the core of the network.

This is commoditized hardware, although the algorithms needed to ensure that servers have the right data on them and would provide vendors with some potential differentiation opportunities. Satellites could potentially be a cost-effective option for provisioning these terrestrial edges of network server platforms.

8.11 Summary

The mobile broadband industry has a two-order-of-magnitude scale advantage over the satellite industry. This scale advantage translates directly into an ability to invest in optimized mass market production and manufacturing techniques.

The satellite industry is consummately good at producing products of the very highest quality but at high cost. Mobile broadband industry scale means that it can deliver quality at low cost with a contemporary smartphone being an exemplary example.

It is possible that the smartphone will become a less important part of the mobile broadband value chain but as yet there is no great sign of this happening and the high expectations for the IoT market have yet to be realized in both volume and value terms.

It is the stated ambition of many players in the satellite industry to deliver far more than just a bit more backhaul. The NEWLEO entities want to reinvent connectivity, but it is hard to see how this will be achieved without a pervasive presence in next generation smartphones and wearable devices such as the Apple Watch and without structuring a commercial ecosystem where both the 5G and satellite community gain from working together.

References

[1] http://www.baesystems.com/en-uk/heritage/british-aerospace-uk.

[2] https://www.isixsigma.com/new-to-six-sigma/history/history-six-sigma/.

[3] OneWeb | OneWorld, "OneWeb Satellites Breaks Ground On The World's First State-Of-The-Art High-Volume Satellite Manufacturing Facility," 2017, http://www.oneweb.world/press-releases/2017/oneweb-satellites-breaks-ground-on-the-worlds-first-state-of-the-art-high-volume-satellite-manufacturing-facility.

[4] http://www.murata.com/en-eu/products/capacitor/mlcc/packaging/pockettaping.

[5] http://pwcircuits.co.uk/fr4.html.

[6] http://www.rockwellautomation.com/global/industries/semiconductor/overview.page.

[7] https://www.bloomberg.com/research/stocks/private/snapshot.asp?privcapId=155172.

[8] https://www.broadcom.com/.

[9] http://www.qorvo.com/.

[10] http://www.twi-global.com/capabilities/joining-technologies/friction-processes/friction-stir-welding/.

[11] https://www.delphi.com/manufacturers/auto/safety/active/electronically-scanning-radar.

[12] https://www.statista.com/.

[13] https://www.gsma.com/IoT/embedded-sim/.

[14] https://www.apple.com/uk/newsroom/2017/09/apple-watch-series-3-features-built-in-cellular-and-more/.

[15] https://www.murata.com/en-eu/products/substrate/ltcc/outline.

[16] https://www.gsmarena.com/asus_zenfone_4_pro_zs551kl-8783.php.

9

Commercial Innovation

9.1 Introduction

Throughout this book, we have been arguing the case for collaboration between the 5G mobile broadband community and national, regional, and sovereign satellite operators. While fine in theory, potential collaboration, even if perceived useful, is frustrated by an adversarial spectrum auction and allocation process. This can only be resolved if both parties recognize that they have a problem to solve and that the other party is part of the solution.

Other third parties and their supply chains also have a role to play, for example, the automotive industry. Automotive manufacturers need to add value to their products, and connectivity is part of the answer. In return, they offer scale by volume and value. Ford sells over 6 million cars per year. Theoretically at least, every car could function as a base terrestrial station and be connected by satellite. Our task in this chapter is to explore how commercial innovation can create this type of scale opportunity.

9.2 The Problem That the Satellite Industry Needs to Solve: A Lack of Scale

We finished Chapter 8 with a brief review of the Apple3 Watch, making the point that products that set new benchmarks for functionality in a small-form factor require market scale to be viable. Scale viability can be estimated in terms of customer reach, cash resources, or borrowing capability, or some combination of all three.

9.3 The Double Dozen Rule

As a rough rule of thumb, scale viability in the global smartphone and 4G and 5G infrastructure industry requires an annual research and development budget of the order of $12 billion to $14 billion. This is typically 12% to 14% of the turnover of companies such as Samsung or Apple. We call this the double dozen rule. The 4G and 5G base station and infrastructure business has a similar scale. Ericsson, Nokia, and Huawei all have research and development budgets of the order of $12 billion to $14 billion per year, which is 12% to 14% of their revenue.

Being in the double dozen club does not guarantee success. In Chapter 8, we referenced the investment by Google of over $12 billion in Motorola Mobility with the business sold to Lenovo 3 years later for $3 billion. Being the world's favourite search engine has not, to date, translated into smartphone market success. HTC, the company from which Google bought the team that designed and manufactured its first smartphone, was in the double dozen club, but its own smartphone market share dropped from 12% to 1%. Intel invested at least $12 billion attempting to buy market share in the LTE baseband business. Broadcom tried a similar strategy and ended up merging its RF assets with Avago, a much smaller company.

Joining the double dozen club is also expensive. Tesla is spending more on research and development and manufacturing investment than they are earning from their combined automotive and energy businesses. This is only possible because the venture capitalist and investor community believe that the company has sufficient future earnings potential to provide a return on the risk.

9.4 National, Regional, and Global Operator and National, Regional, or Global Scale

Self-evidently, with 4 billion customers, the mobile industry does not have a scale issue, but with over 600 individual operators worldwide, it is inevitable that some of them or even many of them are subscale; indeed, MOWO Global [1] believes that only the top 10 operators in the world are scale-efficient. Scale-efficient operators do not necessarily have to be global operators. AT&T is a national operator with limited holdings in rest of the world markets, but its home market has sufficient scale by volume and value to provide a comfortable profit base. AT&T is also profitably invested in fiber, copper, and cable assets. Telstra in Australia is in a similarly fortunate position. The United States and Australia are high average revenue per user (ARPU) markets. AT&T in the United States and Telstra in Australia both have an EBITDA well above the global industry average. The National Broadband Network, operated by Telstra but owned and financed as a nationalized asset, provides a cost-efficient mix of LTE mobile and fixed wireless and fiber with deep rural coverage and some backhaul provided by two geostationary high throughput satellites. The GSM network was turned off in 2017.

China Mobile is also an example of a national operator with a local market large enough for it to be the world's largest operator by volume and one of the largest operators by value. There are also successful although highly geared operators such as Telefonica that started off as regional operators (Latin America and Spain) and then expanded to other markets, including for Telefonica, the United Kingdom.

Going in the other direction, Vodafone started in the United Kingdom in 1985 as Racal Vodafone to provide competition for BT Cellnet. In the heady days of the 1990s, Vodafone's share price soared, but the tiger of the stock market needed to be fed with year-on-year growth, which could only be achieved by aggressive overseas expansion. Thus, Vodafone became arguably the first truly global mobile cellular operator. Many of the national operations retain substantial management and financial independence, although research and development and global strategy are overseen from the United Kingdom. Other operators also have substantial holdings in markets overseas. These are reported in their corporate annual returns as proportionate subscribers. More recently, newer companies such as Digicel have moved from being a national operator to regional to aspiring global operator.

Whether this is the start of a trend towards more global companies running what looks to the user or corporate customer like a global network is probably going to be determined by regulatory and competition policy. In the 1990s, Professor Martin Cave at the University of Manchester developed the theory that the value of spectrum would be maximized if five operators were allowed and encouraged to engage in the bid process. This was a popular theory with regulators and widely adopted throughout the world. It was less popular with operators, and in many markets the five-operator model has proved to be commercially and technically inefficient. This is particularly the case for smaller operators that come into markets years after the incumbents. Companies already operating networks had often started life as a monopoly national operator or been part of a long-term duopoly. The U.S. market is an example. These established operators have customer assets, site assets, and backhaul assets including fiber, cable, and copper in the ground and above ground. Regulators can and do legislate to try and enforce fair access, for example, to fiber, but this is often a less than perfect process. The problem becomes more acute as terrestrial networks densify. Hardware costs reduce, but the cost of digging holes in the ground remains constant. This is an opportunity for satellite operators, although by implication it means they are engaging with the smaller operators in each market.

9.5 The Impact of Standards on Commercial Innovation

In the 1990s, the emergence of GSM as an increasingly dominant global standard meant that global operators could benefit from global scale. Alternative technologies continued to be promoted during and after the introduction of 3G, most notably Wi-Max, and were modestly successful mainly because 3G had underlying power efficiency and performance issues. LTE, first introduced in 2009, has had its critics and some would argue wrongly prioritized spectral efficiency over power efficiency. However, it has become almost completely dominant across the world and now works rather well. Its success has been consolidated by the relentless rise of the smartphone as the companion of choice for billions of subscribers. Standards are critical to the process of commercial innovation and cost reduction including the active management of intellectual property costs through Fair, Reasonable, and Nondiscriminatory (FRAND) agreements, the process of arbitrating intellectual property disputes. The amount of man-hours spent in standards meetings is awesomely large, bigger even than the number of man-hours spent on spectrum issues. The satellite industry has nothing remotely similar in scale. The standardization of the SIM 30 years ago provided the foundation for the fabulously profitable global roaming industry. Standards delivered the market scale to make smartphones viable. Standards have provided the framework on which modern app stores are built and cloud computing is delivered. Standards are enabling operators to develop new markets such as emergency service provision and industrial and consumer IoT.

9.6 Do Mobile Operators Have Any Problems They Need Solving?

Overall, it could be said that mobile operators have spent the last 30 years managing rather well. They have managed to implement four technically complex standards and oversee a spectrum allocation and auction process that is byzantine in its complexity and developed a business that has grown from no subscribers to 4 billion subscribers, many with multiple devices.

9.6.1 Backhaul Costs, Public Safety, and Deep Rural and Desert Coverage

However, there are problems emerging where the satellite industry could be the solution or at least a part of the solution. We have already mentioned backhaul. Mr. Musk, never a man to be underestimated, believes that the SpaceX constellation could deliver 50% of all terrestrial backhaul communications traffic and up to 10% of local internet traffic in high-density cities. OneWeb stated that it is confident that it can substantially reduce 5G backhaul costs both in dense urban and deep rural areas and provide more cost-effective mobile and fixed broadband geographic coverage for rural connectivity. This includes IoT connectivity and developing market connectivity where base station electricity is particularly expensive. These are statements of intent rather than reality but provide an insight into the ambition and fiscal need of the NEWLEOs to couple with the cellular industry more closely.

We have also suggested that, counterintuitively, satellite systems, particularly LEO satellite systems with intersatellite switching, can deliver long-distance, end-to-end latency gain, but the biggest opportunities are probably determined by the fact that many of the emerging applications in 4G and 5G require geographic rather than demographic coverage.

AT&T and their FirstNet contract provide a contemporary example. This is the first nationwide foray by AT&T into providing services to the public safety sector. The traditional two-way radio industry has a 70-year history of servicing mobile radio connectivity to emergency services such as the police, fire brigade, ambulance, and first responder public protection and disaster relief agencies but has struggled to keep pace with LTE device and network functionality. This is because it is a relatively small market with at least three competing technical standards. AT&T have been gifted 20 MHz of 700-MHz spectrum and draw-down rights on $6 billion of funding to deliver the geographic data and voice reach needed for the first responder market in the United States. BT and EE are structuring a similar deal in the United Kingdom, and this looks likely to become a default approach to next-generation public safety and protection mobile connectivity.

Whether this is an opportunity or a challenge for the satellite industry is open to debate. AT&T and any other operator with emergency service sector ambitions will need to invest substantial amounts of money in new sites to meet the geographic service obligations of these new contracts. While it is unlikely that satellite functionality will be added in to LTE 700-MHz FirstNet smartphones, it might make economic sense to provide LTE compatible satellite connectivity to emergency service vehicles.

This amounts to many thousands of vehicles and highlights the fact that satellites are often a more effective and efficient option for connecting objects that move particularly objects that move quickly, which includes trains, boats, and planes. Bear in mind that there are 11,000 registered passenger planes in the world (very hard to service from a 4G or 5G terrestrial network), 50,000 registered merchant ships (impossible to service from a terrestrial 4G or 5G network), and 1.5 billion cars in the world (more efficiently connected via a satellite network particularly from high inclination angles). No one seems to know how many trains there are in the world. It is possible to connect a train traveling at up to 500 km per hour with an LTE network but not particularly efficient. The handover rate from a satellite network will be lower and, in many cases, the link will be line of sight, for example, down into railway cuttings.

There are also 7.5 billion people in the world, which, given that there are only 4 billion cellular phone users, suggests there are 3.5 billion people who are living without a smartphone. However, many of them are living at close to subsistence in deep rural areas and the deserts of the world. This is the geographic scale problem compounded by the demographic scale problem. A lot of countries have a lot of empty space where few people live and the ones who do live there do not earn or own very much. This paragraph, for example, was written on a plane 2 hours from Sydney having spent the last 2 hours flying over more or less nothing else but desert, and Australia is a small country (4,000 km west to east) compared, for example, to Africa (8,000 km north to south).

The African subcontinent is enormous, easily swallowing the United States, China, India, Eastern Europe, France, and Spain, and it has very little fiber. Device costs and network costs must be at least two orders of magnitude lower to make services affordable for people living in these large, empty places.

9.6.2 The Deep Rural Network, Device Cost Issue, and Satellite Solution

In effect, this means producing a smartphone for the cost of a transistor radio. Even if Softbank persuaded ARM to persuade its customers to give their chips away for free, it is hard to see how this could be realized, even for a simple Wi-Fi-enabled device.

The answer is to find another reason to provide connectivity. Figure 9.1 shows what is effectively a Coca-Cola vending machine on wheels and at least partly explains why Coca-Cola, a $40 billion turnover corporation with an enterprise value of $270 billion, was an early investor in the OneWeb consortium.

Figure 9.1 The Coca-Cola Network.

The project is linked to the Coca-Cola 5 by 20 campaign, which aims to create 5 million women entrepreneurs by 2020 [2]. In October 2015, Facebook and Eutelsat made an agreement to provide Wi-Fi to Africa via the AMOS Ka-band GSO satellite as part of Mark Zuckerberg's internet.org initiative. In practice, there was a launch failure [3] and the satellite was destroyed on the launch pad, but Mark Zuckerberg, the founder of Facebook, remains committed to developing low-cost world connectivity projects [4]. This can only happen if device pricing drops to $30 with monthly connectivity costs at a similar level. The connectivity cost is shared between multiple users on the local Wi-Fi. Progress with a pilot project in Uganda suggests that this is achievable [5]. Zuckerberg has announced similar plans for Indonesia.

9.6.3 Low-Cost IoT: Can Satellite Deliver?

This issue of device and service cost translates across into the industrial and consumer IoT market. In an earlier chapter, we referenced Orbcomm as an example of very high frequency (VHF) satellite IoT connectivity coupled with an Inmarsat L-band and cellular modem. However, their users are large, heavy, Earth-moving machines and ships, the Internet of large, expensive, slow-moving objects. The NEWLEO players talk of IoT as a major market, but they are competing with long-range, subgigahertz, narrowband, terrestrial modems and with terrestrial LTE.

Chinese vendors are crashing the cost and price of 4G LTE devices and network hardware. These cost levels are being matched (reluctantly and with some difficulty) by Ericsson and Nokia. The result is that Verizon can introduce sub $30 Cat 1 LTE modems coupled to $2 dollar per month access costs and have stated a longer-term ambition to reduce device cost down towards $3 rather than $30. Whether satellite operators would be willing or able to match these price levels is open to debate. The difficulty is that you cannot charge one set of customers $2 per month for IoT connectivity and another set of customers $200 or $2,000 per month for the same thing. The same tension exists between the legacy SATs and NEWSATs. The legacy SATs have borrowing ratios based on high margins. Reducing margins in anticipation of volume gains that may or not materialize is not intrinsically attractive.

9.7 The CondoSat as an Agency of Change

The constraints of legacy business models do not apply to new market entrants, particularly new entrants that can leverage access to wholesale bandwidth to deliver low-cost access for geographically specific markets. Satellites that provide bandwidth to third parties are known as CondoSats (as in Condominium SATs).

Digicel, for example, started life as a Caribbean operator buying bandwidth from GSO satellites to provide low-cost, wireless fiber connectivity (communications and entertainment) to the Caribbean islands. They have a 2030 transformation program [6], which scales this model from regional to global coverage with an initial expansion into Papua New Guinea.

Buying GSO bandwidth constrains the expansion model in terms of high latitude coverage. Digicel therefore has a mix of GSO and MEO bandwidth; for example, present provisioning for Papua New Guinea includes bandwidth on the SES/O3B MEO constellation. The advantage is that bandwidth contracts can be matched to spot coverage beam footprints so they can be precisely coupled to geographically specific marketing campaigns. Note that 2030 is a useful date as it coincides with the 2015-2030 United Nations Sustainable Development Program [7]. It is helpful for operators and the satellite industry in general to show how their service offers help deliver these goals.

Kacific [8], like Digicel, started life as a lean company (two employees) but has used the CondoSat model to bring connectivity to islands with limited legacy connectivity (Figure 9.2). The company has leveraged their market growth to underwrite their own high throughput satellite with a SpaceX launch booked for late 2018.

The CondoSat business model relies on spare capacity being available and this is dependent on fill ratios, but, to date, high throughput satellite capacity has increased at least as fast as demand. Wholesale-based service offers are therefore a useful agent of change.

9.8 Terrestrial Trash Bin Wi-Fi: Competition or a New Target Market

Sometimes also there may be a local connectivity solution that is more cost-efficient because the site costs are amortized by some other function. An example

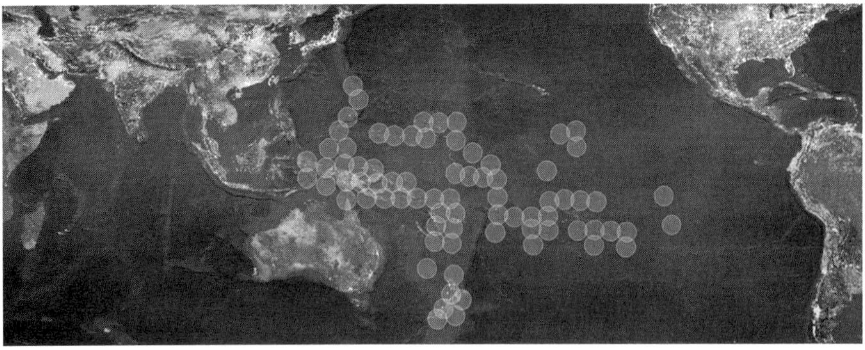

Figure 9.2 Kacific coverage footprint.

from Singapore (a very different market to Australia and the Pacific Islands) is the use of trash bins to provide Wi-Fi from 11 a.m. to 5 p.m. [9] (see Figure 9.3).

9.9 Energy and Carbon Targets: Can Satellite Deliver?

While we are on the topic of recycling and saving the world, can satellites help 5G to meet its energy efficiency and carbon targets? And what are those targets? Hidden deep in the IMT specification documents (Recommendation ITU-R M.2083-0) is a general statement that the energy consumption for 5G should not be larger than present LTE networks. In the NGMN White Paper on energy efficiency [10], the suggestion is that the energy consumption should not be larger than half of existing network consumption but that the network should be capable of supporting a 1,000 times capacity increase. 3GPP wants to study the topic in more detail [11]. Most discussion documents reference small cells as part of the solution, but as already suggested, there is strong evidence that small cells increase rather than decrease energy consumption. The positive start point is that existing networks send most of their RF energy in the wrong direction or backwards. The wrong direction issue can be resolved by using narrow beamwidth antennas and dynamic beamforming, although this can consume significant processing power (although it also reduces system to system interference, so you may have helped someone else meet their energy targets). The issue of energy going backwards can be resolved by improved matching in RF fronts ends, particularly at higher frequencies.

Figure 9.3 Rubbish Wi-Fi in Singapore. (With thanks to Big Belly and Terra Sol Pte Limited.)

9.10 Above the Cloud Computing: Alibaba and Tencent as the Future?

At least we can measure and manage how much damage we are doing to the planet and make some money out of gathering the information, analyzing the information, and selling the information back to the agencies charged with the task of telling us what to do. In an earlier chapter, we referenced Planet.com [12] as an example of a space-based, Earth imaging and sensing platform. We can only manage a problem when we can measure it so the start point is to find an efficient way of delivering the measurement process. In a roundabout way, that brings us to cloud computing and its role as an enabler of commercial innovation in the 5G and satellite industry. In terrestrial networks, vendors are promoting edge computing, which is essentially locating servers in base stations to minimize response latency. Servers in the sky are an alternative option and servers do not need to be stationary, which brings us to the topic of trains, boats, and planes.

9.11 Trains, Boats, and Planes

In an interview with *Satellite Today* in December 2017, Rupert Pearce, CEO of Inmarsat, suggested that Inmarsat's aviation connectivity business, providing while-you-fly Wi-Fi, will be worth $1 billion per year by 2025. This assumes that the number of planes connected to the internet via L-band, S band, and Ka-band Inmarsat GSO satellites will grow from a present installed base of 1,200 planes to 5,000 planes each yielding $200,000 of annual revenue.

9.12 Mobile Automotive Mobile Networks

There are emerging business models where the cost of connectivity is potentially cross-amortized with the value of data collected from Earth-based moving objects. Pirelli, for example, captures information from sensors in a car including wireless tyre pressure monitoring for uploading to the Pirelli cloud. Value is extracted from this data, for example, by sending e-mails to customers when their tires need replacing. Continental has a similar system.

9.13 Satellite and 802.11p Automotive V2V and V2X

Cars of the future also have to chat among themselves apparently and to the 5G automotive network. Vehicle-to-vehicle is called V2V and vehicle-to-network is called V2N, or generically, V2X.

Figure 9.4 is a computer-generated image produced by the National Traffic Safety Association in the United States [13]. The assumption is that V2V and V2X will be managed over dedicated ISM spectrum above the 5-GHz Wi-Fi band using the 802.11p standard, a subset of the 802.11 Wi-Fi standards process.

Most satellite operators will immediately have noticed the absence of any satellite links in the image, even though directly overhead LEOs are significantly better positioned to have an overall view from space of local traffic issues and accident hazards.

There will be a need for local connectivity to meet millisecond latency requirements, but satellites are critical to the delivery of a safe semiautonomous and autonomous transport experience.

9.14 Subgigahertz CubeSat as an Alternative Delivery Option Using Sub-1-GHz Spectrum

Alternative satellite delivery models are emerging based on sub-1-GHz spectrum. Myriota, a start-up CubeSat operator initially servicing Australia, collects data from water tanks and agricultural sensors in the outback (Figure 9.5) using data sensors equipped with a 33-mW transmitter to uplink 20-byte data packets every 90 minutes to a CubeSat as it passes 800 km overhead [14].

Figure 9.4 Satellites and 802.11p Automotive V2V and V2X. (Image courtesy of the NHTSA.)

Figure 9.5 Satellite water tower monitor on a Merino sheep farm in Australian outback. (With thanks to Myriota.)

9.15 Space-Based White Space

Theoretically at least, it might be possible to consider using white space spectrum for satellite connectivity. As a reminder, the principle of white space is that spectrum that is unused either for a certain time or at a certain place is made available to third parties. TV white space, for example, was proposed as a database mechanism for using the channels that were not being used in multifrequency terrestrial broadcasting between 450 MHz and 890 MHz. In practice, TV broadcasting has been packed into single-frequency networks at the lower end of the band and most of the rest of the spectrum has been auctioned off to mobile operators.

9.16 Space and HAPS-Based Wi-Fi

Another alternative is to build a space-based or high altitude platform stations (HAPS)-based Wi-Fi network. The problem with this to date has been that the Wi-Fi physical layer does not scale to long-distance path lengths because it is TDD with relatively short time-domain guard bands. Power outputs are also low, 10 mW for devices and 250 mW for access points. However, the emerging 802.11ax standard includes an FDD option. This is being introduced to provide a way of managing interference in dense indoor Wi-Fi deployments, for example, managing multiple Amazon Echo or Google Home devices, but this also opens the opportunity to deliver longer path length Wi-Fi. This could include Wi-Fi implemented in the subgigahertz ISM bands. The Myriota example referenced above illustrates that an uplink power budget of a few tens of milliwatts is more than adequate for supporting many low data rate IoT direct to space or direct to HAPS systems, the Google Loon project being one example.

9.17 The Smartphone as the Default Common Denominator for B2B and Consumer Mass Markets

It is impossible to ignore the fact that an increasing majority of B2B and consumer mass market applications are delivered over a smartphone. Data collected from the Myriota CubeSat is placed on a server that is accessed over a Wi-Fi or LTE connection, but as far as the farmer or vineyard owner is concerned, it is his or her smartphone that is doing the hard lifting.

This raises the question as to whether the satellite industry should be working towards embedding direct rather than indirect connectivity into 5G smartphones, to be the pipe rather than part of the pipe.

9.18 The 5G Smartphone as the Gateway to Satellite Industry Consumer Market Scale

It is now clear from vendor briefings what a 5G smartphone will look like. It will be introduced initially into the U.S. market and will be a 4G phone (the initial 5G networks will be nonstand-alone, coupled to the LTE control plane) with a 28-GHz 5G radio coupled to a 16-element or 32-element active electronically steerable array antenna printed into the smartphone display, which will also be acting as a solar panel.

The antenna array will be optimized to work in the vertical plane, but if you place the phone on a flat surface it will be looking upwards at LEO, MEO, and GSO satellites.

This is the single and most compelling reason for supporting cosharing of the 28-GHz band between high throughput satellite and terrestrial 5G. There will be minimal motivation for smartphone manufacturers to add satellite-only bands to a 5G phone. There is a compelling motivation to support in-band satellite connectivity.

For the 5G mobile operator and their customers, it adds no cost to the smartphone and it means that the 5G smartphone can be used outdoors anywhere in the world including deserts, oceans, and areas where geographic coverage is hard or impossible to deliver economically from terrestrial networks.

The satellite operator provides the opportunity for the first time in the industry's history to capture mass market consumer added value. The positive impact on the EBITDA and enterprise value of 5G mobile operators and their new best friends in the satellite industry will be spectacular.

9.19 Wireless Wearables

To reprise our momentary flight of fancy into the wearables market (see Chapter 6), it would seem plausible to embed terrestrial and satellite connectivity into clothing for professional use and leisure use with antennas on the front, back, and sleeves of next-generation outdoor clothing. The combination of metamaterials and electronic bandgap materials (Chapter 6) means that low-count and high-count element passive antennas could be woven into the fabric of these clothes and provide connectivity that could scale from VHF LEO satellite constellations through UHF (Band 31) and L-band, S-band, and C-band to the K-bands and V-band and W-band (from 138 MHz to 92 GHz). The clothing-based antennas would act as high gain antennas for body-worn devices including monitoring and communication devices, offering the opportunity to develop added value applications beyond present wireless wearable product offerings.

9.20 Back to the Beach in Bournemouth

In Chapter 1, we provided an example of the beach in Bournemouth as an analogy of how value can be derived from monitoring how we interact with the physical world around us. I wrote this paragraph in a kitchen in my cousin's house in Sydney and I uploaded the text to somewhere in the cloud over Wi-Fi, although it might have been LTE, but in the process, someone somewhere knew where I was and what I was doing. The general point to make is that commercial innovation in the 5G and satellite industry is not just about connectivity but the control of that connectivity.

9.21 Getting 28-GHz Satellite Connectivity into 5G Smartphones: The Practicalities

The idea that I can be connected and in control anywhere in the world including all those not spots where my phone struggles to find a signal is a compelling proposition. However, there are practical issues to address including realizing a low-cost, power-efficient, 28-GHz front end within a smartphone form factor, which is already full of other radio components.

9.22 Getting C-Band (and Extended C-Band), S-Band, L-Band, and Subgigahertz Satellite Connectivity into Smartphones

Having sorted a 28-GHz satellite front end and baseband for 5G phones, the next challenge is to get agreement to support satellite connectivity into 4G and 5G spectrum in C-band (3.4 to 3.8 GHz) and the core LTE and 5G refarming bands, Band 7 (2.6 GHz), Band 1 (1.9/2.1 GHz), 1.8-GHz and 1.9-GHz PCS, and the subgigahertz bands, bearing in mind that satellite operators have existing spectral assets adjacent to some of these bands. Bands 1 and 7, for example, could have extended passbands by combining terrestrial and satellite allocations.

9.23 Standards as a Critical Enabler

The idea that the satellite industry should welcome the 5G community into their Ku-band and Ka-band spectrum might seem hard to swallow but the bitter pill would be substantially sweetened if reciprocal access rights into the sub-3.8-GHz LTE and 5G bands could be agreed.

If this could be coupled to the default inclusion of in-band satellite connectivity into smartphones and wearable devices, then the satellite industry would have achieved access to consumer market scale and connected consumer added value. However, smartphone manufacturers need to be motivated to add satellite connectivity. The best way to do this is to make sure that additional value is realizable at no extra cost.

This means that in-band cosharing is necessary and needs to be combined with a satellite physical layer that is compatible with existing and future smartphone and 5G IoT RF front ends. The complex modulated waveforms from a satellite transmission must sit comfortably within existing and future 4G and 5G channels within existing and future 4G and 5G passbands and be able to be processed through existing and future 4G and 5G switch paths and filters and RF power amplifiers and low noise amplifiers.

Agreement therefore must be reached on spectrum and standards; the choice of physical layer is as important as the band plan. This requires effective engagement between the satellite industry and 5G standards process.

References

[1] https://www.mowo.global/.

[2] http://www.coca-colacompany.com/5by20.

[3] https://spaceflightnow.com/2016/09/01/spacex-rocket-and-israeli-satellite-destroyed-in-launch-pad-explosion/.

[4] https://info.internet.org/en/.

[5] https://techcrunch.com/2017/02/27/facebook-digs-into-mobile-infrastructure-in-uganda-as-tip-commits-170m-to-startups/.

[6] https://www.digicelgroup.com/en/about/digicel-2030.html.

[7] http://www.un.org/sustainabledevelopment/sustainable-development-goals/.

[8] http://kacific.com/the-project/.

[9] http://www.eco-business.com/news/turning-trash-bins-into-hotspots-for-info-and-money/.

[10] https://www.ngmn.org/5g-white-paper.html.

[11] 3GPP TR 38.913 (section 7.19).

[12] https://www.planet.com/.

[13] https://www.nhtsa.gov/.

[14] http://myriota.com/.

10

Standards

10.1 Standards as a Barrier to 5G Satellite Smartphones

In Chapter 9, we argued that adding satellite connectivity to smartphones would substantially improve the consumer experience, adding value by delivering global coverage. This, in turn, would increase the EBITDA and enterprise value of the mobile operator community and satellite community. The connectivity would initially be in the 28-GHz high throughput passbands but then scale into the sub-3.8-GHz LTE bands including the core bands likely to be used for 5G refarming.

However, the mobile operator community has weaponized the standards process with Releases 15 and 16 closely coupled to a campaign for primary access rights to satellite high throughput spectrum at WRC 2019. Releases 17 and 18 similarly will be used as the basis for campaigning for primary access rights to V-band VHTS spectrum and E-band super VHTS spectrum.

Conversely, satellite operators have adopted a "get your tanks off my lawn" attitude to their existing high throughput spectral assets in Ku-band (12 GHz) and 28 GHz (Ka-band). This is a process that will destroy longer-term enterprise value for all parties involved.

10.2 Standards as an Enabler of 5G Satellite Smartphones

Changing the polarity of this process from negative to positive, standards can be regarded as essential part of a coupling process between two entities presently engaged in mortal combat but who should be embracing one another in a standards lovefest.

One difficulty is that satellite operators have very lean human resources. Many functions are outsourced and the revenue per employee when compared to mobile operators is astronomically high; over $30 million per employee is not uncommon.

This means that sending satellite operator staff to standards meetings incurs eye watering opportunity costs. However, we would suggest that if the satellite industry wants to remain independent, it will need to scale into consumer markets. Scaling into consumer markets implies a need to capture value from smartphones and other emerging consumer wireless markets including wireless wearables. This can only be done by finding a way of coupling with the 5G standards process and related local area and personal area connectivity standards.

A starting point is to understand how the mobile broadband standards process works by studying the internal and external tension points.

10.3 The Use and Abuse of the Standards Process: Internal Tension Points

A recent blog post from the vice president of technical standards at Qualcomm complained that the 3GPP standards process was being manipulated by other participating companies exploiting an overly simplistic contribution count system [1]. Some might say that this is an example of the pot calling the kettle black, but the blog makes useful points about the need to improve the present standards process. In this chapter, we explore the inherent disconnects between standards making, spectrum allocation, auction policy, and competition policy and suggest that an adversarial approach to the repurposing of spectrum and related changes to spectral access rights is not a good basis for standards integration, but finding a better approach is not easy.

For vendors, the incentive for participating in standards groups is that 3GPP members can seek intellectual property rights in accordance with the intellectual property rights (IPR) policies of the regional standards setting authorities, the European Telecommunication Standards Institute (ETSI), the Association of Radio Industries and Businesses (ARIB) in Japan, the Alliance for Telecommunication Solutions (ATIS) in the United States, the China Communications Standards Association (CCIS), the Telecommunications Standards Development Society (TSDI) in India, the Telecommunications Technology Association (TTA) in Korea, and the Telecommunications Technology Committee (TTC) in Japan.

While this is understandable, it must be remembered that the purpose of a standard is to realize market efficiency by facilitating interoperability and market scale. In communication systems, interoperability and market scale are dependent on spectral harmonization.

The harmonization process and the standardization process should be consensus based but in practice are influenced by special interests. These special interests can be region or country-specific and the differences can be subtle but significant, wider passbands or different out-of-band requirements, for example, mean that either scale benefit has to be sacrificed or radio frequency (RF) hardware has to be characterized for the worst-case conditions, in this example, the highest protection ratios. This will have an impact on device and network cost and performance.

The present structure used for 4G and 5G standards dates from 1998 with the formation of 3GPP, partly driven by the recognition that the United States and rest of the world cellular standards needed to be brought together. Release 99 was the first standard issued by 3GPP with the intention that future release dates would happen yearly. Given that we are now on Release 15, this has not quite happened, but the principle still applies. Release 15 is the first release to specifically address 5G physical layer standards and upper stack optimization.

Twenty years of 3GPP standards have had to couple into 150 years of spectrum policy making under the auspices of the ITU. The ITU divides the world into three regions, Region 1 for Europe and Africa, Region 2 for America and Latin America, and Region 3 for the Asia Pacific and Australia. Historically, this has encouraged regional-specific standards to be deployed into region or country-specific spectrum, the Personal Digital Cellular Standard in Japan at 1.5 GHz and IS95 CDMA and IS54 and IS136 TDMA in the U.S. 800-MHz band were two examples.

Legacy allocation decisions taken on a regional basis, for example, the allocation of an ISM band between 902 and 928 MHz in the United States continues to influence band plans and explains why the United States does not have any 900-MHz cellular networks. An apparently minor regulatory decision can have a major long-term impact. A cellular band at 800 MHz in the United States means that the whole sub-1-GHz band plan is different to the rest of the world.

However, spectral access rights are ultimately a sovereign responsibility. Every nation has a right to the final say on how spectrum will be used within its borders provided that coexistence with other geographically adjacent countries meets internationally agreed criteria.

In practice, scale economies dictate that counties chose to harmonize their spectrum band plans regionally and when possible globally. There are also operator-specific requirements. These have become increasingly complex due to the perceived need to support channel aggregation.

Dish Networks Band 70 is an example from the United States, a concatenation of their AWS 4 spectrum (2,000–2,020 MHz), their H Block PCS spectrum (1,995–2,000 MHz), and unpaired AWS-3 spectrum (1,695–1,710 MHz). Sprint is another U.S. operator-specific example with their Gigabit LTE tri-band proposal combining their 800-MHz, 1,900-MHz, and 2.5-GHz band allocations. 3GPP addresses these regional, country, and operator-specific requirements by producing technical specifications; a specification, as the word implies, is specific to a particular requirement.

In 5G an additional level of complexity is introduced by the need to accommodate vertical markets. This is broadly covered by developing work streams for different requirements, for example, enhanced mobile broadband (eMBB), massive machine type communications (mMTC), and ultrareliable, low-latency communication (URLLC).

In practice, particular industries are going to have particular requirements that will need to be met. 3GPP has to work with parallel standards making organizations including IEEE and higher-layer protocol standards bodies such as the Internet Engineering Task Force (IETF) developing vertical market-specific profiles with the vertical market standards bodies. Utilities, for example, have different standards in different countries; even countries within the EU can have marked differences in the way that electricity, water, and gas are managed, monitored, measured, and regulated.

The ongoing work to develop a 5G automotive industry offer is another example. Automotive industry standards are at least as complex as telecommunication standards and have multiple touch points with IEEE standards making including 802.15.4 and 802.11-based connectivity. Specifically, work outputs from 5GAA (the 5G automotive association) will need to be closely coupled with IEEE 802.11p standards and spectrum band plans.

This is made harder by the move within 3GPP to introduce licensed spectrum standards into unlicensed spectrum (LTE-U and LTA Licensed Assisted Access). Coexistence issues, whether real or imagined, are not a good basis for constructive standards engagement.

However, there will also be a need to integrate 5G vertical market work items with vertical market work outputs from other parts of the telecommunications supply chain including the satellite industry. The announcement that the nonstand-alone (NSA) implementation of the 5G New Radio (NR) physical layer will be complete by the end of this year with large-scale trials and deployments in 2019 suggests an ambition that will not be welcomed by the existing satellite operator incumbents in the target bands (3.5 GHz, 4.5 GHz, 28 GHz, and 39 GHz).

This brings us to the thorny question of competition policy. The purpose of competition policy or the related discipline of antitrust policy is to counter monopolistic behavior and to ensure efficient markets.

Antitrust legal cases can take years to resolve. Intel is still fighting a $1 billion fine imposed 7 years ago by the European Commission for alleged anticompetitive behavior against AMD. Qualcomm has been facing resistance from the European Commission to their proposed takeover of NXP.

The mobile operators are additionally constrained by auction policies, which are country-specific, but which have generally followed the principle that five operators per market produce the most market-effective, although not necessarily the most cost-effective, outcome. In practice, deploying multiple parallel networks can be ludicrously wasteful and particularly expensive for market entrants who do not have existing fiber and site assets.

The standard process in its own right could be considered anticompetitive because it makes market entry disproportionately expensive, a lesson that Intel and Broadcom learned with LTE.

However, it is easy to identify weakness in existing practices and processes but hard to suggest better alternatives. To quote Mr. Churchill, "Democracy is the worst form of government except for all the others," and it may be that our existing standards and spectrum policy making procedures are as good as they are going to get.

With the words of Mr. Churchill ringing in our ears, let us move on to the magical world of 5G and satellite standards.

10.4 5G and Satellite 3GPP Release 15 Work Items

There have been several unsuccessful attempts to develop integrated mobile broadband and satellite standards, for example, in 3G with the S-UMTS standard [2]. There have also been attempts to standardize hybrid terrestrial and satellite connectivity through the Auxiliary Terrestrial Component Specifications [3] in the United States, Canada, Europe, and Asia and, in China, the Satellite and Terrestrial Multi Service Infrastructure [4].

At a 3GPP Technical Standards (TSG) Group meeting in March 2017, it was agreed that a 5G and nonterrestrial networks (NTN) study would be produced within the 3GPP Release 15 standards process (New Radio NTN, NR.NTN) [5]. The sponsors include Thales, Dish Networks, Fraunhofer, Hughes, Inmarsat, Ligado, Motorola, Sepura (emergency service radio), the Indian Institute of Technology, Avanti, Mitsubishi, China Mobile, and the Airbus Group.

A list of sponsors is not a guarantee of future progress, but at least a minimum of progress has been made. The relevant standards references are linked to use cases as listed here:

- The support of 5G connectivity via satellite within 3GPP TR23.799;

- The higher availability requirement within 3GPP TR22.862;
- The wide area connectivity requirement within 3GPP TR22.863;
- The satellite access requirements within 3GPP TR 22.864;
- The 5G connectivity using satellites use case of 3GPP TR 22.891;
- The satellite extension to terrestrial within 3GPP TR 38.913.

The definition of a nonterrestrial network is a network or segment of networks using an airborne or spaceborne vehicle for transmission. Spaceborne vehicles include LEO, MEO, and GSO satellites and highly elliptical orbiting (HEO) satellites. Airborne vehicles include unmanned aircraft systems (UAS), tethered UAS (blimps), lighter than air UAS (LTA), heavier than air UAS (HTA), and high-altitude UAS platforms (HAPs).

The statement of work states the desired outcome as:

- Enabling ubiquitous 5G service to UEs (especially IoT/MTC, public safety/critical communications) by extending the reach of terrestrial-based 5G networks to areas that cannot be optimally covered by terrestrial 5G networks.
- Enabling 5G service reliability and resiliency due to the reduced vulnerability of airborne/spaceborne vehicles to physical attacks and natural disasters. This is especially of interest to public safety or railway communication systems.
- Enabling connectivity of 5G-RAN elements to allow ubiquitous deployment of 5G terrestrial networks.
- Enabling connectivity and delivery of 5G services to UE onboard airborne vehicles (including air flight passengers, UASs, and drones).
- Enabling connectivity and delivery of 5G services to UE onboard other moving platforms such as vessels and trains.
- Enabling efficient multicast/broadcast delivery of services such as A/V content, group communications, IoT broadcast services, software downloads (for example, to connected cars), and emergency messaging.
- Enabling flexibility in traffic engineering of 5G services between terrestrial and nonterrestrial networks.

The Release 15 work items are divided into two activities:

Activity A:

- Study Physical layer impact through the characterization of the operational conditions of NR in the nonterrestrial networks. Key design requirements will be identified along with possible solutions for an efficient operation of NR.
- Characterize the operational conditions of NR in selected nonterrestrial networks, and identify key design requirements and issues that need to be solved for an efficient operation of NR such as synchronization, initial access, random access, data channels, channel estimation, low PAPR modulation, and link establishment/maintenance, focusing on:
  - *Channel model:* Study whether existing channel models (3GPP or ITU) can be applied for these links and identify/define improved channel model(s) if necessary. In addition to the outdoor-to-outdoor, the study shall include outdoor-to-indoor scenarios (e.g., providing services to UEs inside a ship, train, or building). [RAN1]
  - *Interference:* Nonterrestrial systems have different interference characteristics (intrasystems and intersystems) compared to traditional cellular networks. Thus, one objective of this study is to understand the interference characteristics. [RAN1]
  - *Doppler effects:* Characterize the impact and identify solutions to compensate for Doppler shift and its spread associated with nonterrestrial communication links. [RAN1]
  - *Propagation delays:* Characterize the impact of propagation delay associated with nonterrestrial communication links (nonterrestrial vehicles operate at various altitudes from very low and comparable to terrestrial networks as UAS and HAPs to low and medium-altitude LEO/MEOs as well as high-altitude GEO/HEOs) and identify appropriate solutions. [RAN1]

Activity B:

- Study impact on Layer 2 and above, and RAN architecture based on NR Phase 1 findings and other operational requirements.
- In this activity, requirements related to higher layers will be studied and potential solutions will be identified including analyzing their performance gains. In particular, the following aspects will be studied.
  - *Propagation delay:* Identify solutions related to Layer 2 protocols and timing relationships to support nonterrestrial network propagation delays. [RAN2]

- *Inter-RAT handover:* Study and identify mobility requirements that may be needed for some nonterrestrial vehicles (such as LEO/MEO satellites) that move at much higher speed but over predictable paths. [RAN2]
- *Architecture:* Identify needs for the 5G's Radio Access Network architecture to support nonterrestrial networks. [RAN3]

The study has to be regarded as only the start of a potentially long and difficult journey. Figure 10.1 lists 19 3GPP work groups that would need to be included in order to realize a comprehensive implementable and testable global standard.

Additionally, the group only started work in October 2017.

The satellite industry does not have a legacy of robust standards making and has a history of implementing a range of different system specific proprietary air interfaces that are only compatible at the higher layers of the protocol stack. This frustrates potential economies of scale particularly in terms of RF hardware compatibility. The satellite industry is also leanly resourced and does not have thousands of engineers available to engage in the 3GPP standards process.

10.5 Parallel Guided Media Standards

Mobile broadband operators and their vendors and satellite operators and their vendors assure us that our wireless connectivity experience will be similar and sometimes better than our wireline connectivity experience.

In practice, the wireline connectivity experience is steadily improving and represents a moving target which wireless needs to track. This implies a need to keep an eye on copper, cable, and fiber standards. This includes standards such as DOCSIS 3.0 and 3.1, vectored VDSL and G. fast variants, and the recently announced MoCA Access. So copper still counts, and it is getting better or rather we can access more channel bandwidth by working at higher frequencies, for example, 8.5 MHz, 17.7 MHz, and 35.33 MHz and with higher-order modulation, for example, 1,024 or 4,096-level QAM. DOCSIS 3.1 notably reintroduces FDD into the media multiplex.

10.6 5G, Satellite, and Fixed Wireless Access

Some markets have deployments of fixed wireless between 3.4 and 3.8 GHz using Wi-Max TDD equipment (from manufacturers such as Motorola that became heavily invested in the standard 15 years ago). Australia is one example

TSG RAN Radio Access Network	TSG SA Service & Systems Aspects	TSG CT Core Network & Terminals
RAN WG1 Radio Layer 1 spec	SA WG1 Services	CT WG1 MM/CC/SM (Iu)
RAN WG2 Radio Layer 2 spec Radio Layer 3 RR spec	SA WG2 Architecture	CT WG3 Interworking with external networks
RAN WG3 Iub spec, Iur spec, Iu spec UTRAN O&M requirements	SA WG3 Security	CT WG4 MAP/GTP/BCH/SS
RAN WG4 Radio Performance Protocol aspects	SA WG4 Codec	CT WG6 Smart Card Application Aspects
RAN WG5 Mobile Terminal Conformance Testing	SA WG5 Telecom Management	
RAN WG6 Legacy RAN radio and protocol	SA WG6 Mission-critical applications	

Figure 10.1 3GPP work groups. (With thanks to 3GPP.)

market where mining companies use this band /technology combination to provide high bit rate connectivity for localized operations. Citizens Broadband Radio and the WISPS (Wireless Internet Service Providers) in the United States also use this band. The Australian and U.S. operator and user communities are understandably keen to protect their access rights to this spectrum. The 5G community are keen to deploy 5G into this band due to the 400 MHz of contiguous bandwidth available.

10.7 5G, Satellite, and C-Band Satellite TV Standards

C-band satellite TV at 3.8–4.2 GHz remains important in many markets coupled to high-definition satellite TV at 12 GHz and superhigh definition at 18 GHz. This is important in terms of interference protection ratios which in turn are influenced by codec standards. High-order codecs, designed to maximize throughput through satellite broadcasting passbands, can be susceptible to interference induced error extension, which can compromise voice and picture quality.

Satellite TV operators such as Dish Networks in the United States have ambitions to transition their satellite TV networks to a two-way high throughput service offer, although this requires a constellation upgrade and changes to regulatory licenses.

10.8 5G and Satellite Integration with the Wi-Fi Standards Process

10.8.1 SAT-FI

Most of the NEWLEO (OneWeb, SpaceX, LeoSat) and NEWLEGACY LEO, MEO, and GSO operators include Wi-Fi as an integral part of their service offer. The Coca-Cola vendor store with an integrated OneWeb transponder referenced in Chapter 9 is intended to serve up Coca-Cola and Wi-Fi connectivity. At the time of this writing, smartphones were beginning to include 802.11ad, which includes a 60-GHz Wi-Fi transceiver. It can also be expected that future phones will support 802.11ax, which includes FDD Wi-Fi. How this will be integrated into unlicensed spectrum band plans remains an open question yet to be seriously discussed by the regulatory community. As referenced in previous chapters, products such as Amazon Echo and Google Home requiring high-density access points are likely to add urgency to this arcane (but important) standards area.

10.8.2 High Data Rate Wi-Fi, Cat 18 and Cat 19 LTE, and 50X 5G

High data rate Wi-Fi is required to share a limited amount of real estate in a smartphone with high data rate LTE. As of late 2017/2018, Category 18 and Category 19 modems were being announced capable of delivering headline downlink data rates of 1.2 Gbps and 1.6 Gbps, although this is dependent on having an RF front end that can support either four of five 20-MHz carriers (80 or 100 MHz of aggregated bandwidth). A limited number of operators have deployed low-band (sub-1 GHz), mid-band (<2 GHz), and high-band (>2 GHz) aggregated carriers to deliver a 1-Gbps service offer, although phones need to be in ideal propagation conditions with low interference to achieve this. Baseband offers for first-generation 5G smartphones have been announced with road maps to take headline data rates to 10 Gbps, the 5G X factor (10 times faster/more efficient than 4G). Qualcomm with their 50X chip set are an early example. This illustrates the close link between standards, supply chain market push and spectrum allocation and planning and explains the evolving focus on sub-3.8-GHz 5G refarming in addition to 26-GHz and 28-GHz deployments. Initial U.S. deployments are in the 28-GHz band, a band explicitly excluded for 5G in other markets. In Chapter 12, we suggest that there is a potentially compelling technical and commercial case for satellite and 5G (and 5G backhaul) to coshare the 28-GHz band, although this is not reflected in existing ITU discussions or satellite industry positioning.

10.8.3 LTE and Wi-Fi Link Aggregation

There seem to be several schools of thought as to how the Wi-Fi bands will develop in the future. The generic driver is the assumption that a high percentage of traffic, some vendors and operators suggest about 80%, will be consumed indoors.

The options are:

- The use of standalone 802.11 ac/ad/ax/ay at 2.4 GHz, 5 GHz and 60 GHz;
- All of above integrated with LTE in licensed spectrums with a multiplex at IP level using an IP SEC (security tunnel) (LWIP);
- License-assisted access, which aggregates unlicensed spectrum with an anchor in licensed spectrum;
- The use of LTE and 5G in unlicensed spectrum, which includes a product offer from Qualcomm called MulteFire [6].

10.9 5G, Satellite, and Bluetooth

The satellite industry has at least some pre-advice of what terrestrial data rates are likely to be in 3 to 5 years' time given that it takes 3 to 5 years for new modem categories to achieve significant market penetration. However, offered traffic is also influenced by other factors such as Wi-Fi local area availability (as above) and personal area connectivity including Bluetooth.

The publication of Version 5.0 of the Bluetooth specification in December 2016 marked another step in the evolution of Bluetooth as a closely coupled technical and commercial partner to 4G LTE and 802.11 Wi-Fi. Most of us use Bluetooth every day whether it is our fitness tracker talking to our smartphone or hands-free pairing in our car or listening to Spotify via a Bluetooth headset.

5.0 Bluetooth is significant for several reasons. It is the fourth iteration of Bluetooth Low Energy (BLE) first introduced as Version 4.0 in 2010, but it is the first time that BLE has been coupled with a long range (1,600m) Bluetooth PHY and MAC option. This is achieved through a combination of higher output power (+20 dBm/100 mW), channel coding and optimised receiver design taking advantage of a feature called stable modulation index where the deviation of the GFSK deviation is reduced. This enables a range of new applications including long-distance retail beaconing.

Long-distance Bluetooth can also be extended with the newly supported mesh protocol. This brings Bluetooth into direct competition with other radio systems including 802.15, 4 based protocols such as Zigbee, LoRa, Wireless-M (for meter reading), Thread, and 6 LowPAN (IPV6 over local area networks). The 802.11 also has a mesh protocol and long-distance ambitions including 802.11ah Wi-Fi in the 900-MHz ISM band. It also moves Bluetooth into the application space targeted by LTE NB IoT and LTE M though with range limitations.

There are some interesting design challenges implied by 5.0. The BLE specification is inherently less resilient to interference than classic or EDR Bluetooth. This is because the 78 legacy 1-MHz channels within the 20-MHz, 2.4-GHz passband are replaced with 39 2-MHz channels with three fixed nonhopping advertising channels in the middle and edge of the passband.

These must withstand high-power 20-MHz LTE TDD in Band 40 (below the 2.4-GHz passband) and high-power 20-MHz LTE TDD in Band 41 above the passband (and Band 7 LTE FDD). This includes 26-dBm high-power user equipment.

The coexistence of Bluetooth, Wi-Fi, and LTE has been intensively studied and worked on for over 10 years and is now managed with surprising effectiveness within a smartphone through a combination of optimized analog and digital filtering (SAW and FBAR filters) and time-domain interference mitigation based on a set of industry standard wireless coexistence protocols.

The introduction of high-power Bluetooth however implies that this is no longer just a colocation issue but potentially a close location issue. Even managing Bluetooth to Bluetooth coexistence becomes a nontrivial task when you consider that +20 dBm transmissions will be closely proximate to −20 dBm or whisper mode −30 dBm transmissions and RX sensitivity of −93 dBm, potentially a dynamic range of 120 dB. Although Bluetooth is a TDD system, this isolation requirement will be challenging and vulnerable to ISI distortion.

More broadly, there is a need to consider how 5G Bluetooth couples technically and commercially with 5G including 5G IoT and satellite IoT. Superficially, it might be considered that Bluetooth and indeed all 2.4-GHz ISM-based systems would not need to be considered within the 5G standards and product definition process. After all, much of the implementation focus is on Ka-band at 26 GHz or 28 GHz. However, as we have seen with LTE-U, taking over ISM spectrum is always a tempting prospect.

This is particularly true when you consider the recent FCC Notice of Inquiry into repurposing of spectrum adjacent to the 5-GHz ISM band (the reallocation of 5.925–6.425-GHz spectrum) and the addition of substantial spectrum to the 60-GHz unlicensed band.

Arguably, the ISM band at 2.4 GHz is so crowded that it will become increasingly unusable, although somehow it continues to work most of the time. The 2.4-GHz band is book ended by Band 40, Band 41, and Band 7. Sprint, for example, has stated an ambition (backed by closely managed demonstrations) to implement Gigabit LTE based on three 20-MHz aggregated channels at low band (800 MHz), mid band (1,900 MHz), and high band (Band 41).

Pragmatically, it seems more than likely that 5G will become part of mobile operators' future refarming plans for sub-3-GHz spectrum including Band 40 and Band 41. It is also likely that there will be a requirement to deliver high-power 5G user equipment including 1-W and 3-W handsets for public protection and disaster relief, scaling networks such as FirstNet and BT EE Emergency Services Networks (ESN) to higher-bandwidth, higher-frequency, shorter-wavelength spectrum. This is all about flux density and uplink range and satellites can and should be part of this delivery model.

When you consider that existing LTE user equipment specified with a maximum output power of+23 dBm has a peak to envelope power envelope of at least +33 dBm, then it seems obvious that there are likely to be significant coexistence issues between 5G sub-3-GHz radio systems and 2.4-GHz ISM including 5.0 Bluetooth and Wi-Fi in all its alphabetical glory.

It is also quite possible that the Bluetooth SIG could have ambitions to scale future standards (6.0 and beyond) into the 5-GHz and 60-GHz bands. The extended 60-GHz ISM band will be immediately spectrally proximate to the proposed 5G E-band duplex at 71 to 76 GHz and 81 to 86 GHz.

This would suggest a need to qualify the proposed modulation waveform candidates in terms of coexistence with existing and potential future ISM radio systems including Bluetooth as a closely coupled partner in 5G wide area, local area, personal area, and IoT vertical market use cases.

In many instances, these terrestrial coexistence issues could be resolved by delivering traffic directly up and directly down from users and devices to directly overhead LEO, MEO, and GSO satellites. This leads us to our next topic.

10.10 How Satellites Can Help Meet Performance Targets Specified in the 5G Standards Documents

ITU draft document 5/40E/ (February 2017) sets out the four application domains or usage scenarios that determine the upper and lower performance bounds of the 5G physical layer and upper layers, specifically eMBB, low mobility large cell (LMLC), mMTC, massive machine-type communications, and ultra-reliable low-latency communication (URLLC).

10.10.1 eMBB and Satellite

For eMBB, the peak downlink data rate is 20 Gbps with an uplink data rate of 10 Gbps. It is unlikely that any satellites could deliver these data rates to a smartphone optimized to receive signal energy from a horizontal rather than vertical plane but very feasible to deliver these data rates (and potentially higher data rates) to upwards facing active or passive flat panel arrays.

10.10.2 Satellites and 5G Spectral Efficiency

5G downlink spectral efficiency is specified at 30 bit/s/Hz, uplink spectral efficiency is specified at 15 bps Hz (due to the use of lower order modulation to meet power efficiency requirements and lower-order spatial multiplexing).

We argue in Chapter 12 that it would be theoretically possible to support 5G, 5G backhaul, and LEO, MEO, and GSO connectivity in the same passband, for example, in the 28-GHz band, which would deliver system spectral efficiencies of a significantly higher order.

10.10.3 Satellites and 5G Deep Rural IoT

Low-power devices are standardized either as Cat 1 (Category 1) or Category 0.

The physical layer specification is summarized in Table 10.1.

The elapsed time between wake-up calls can be any time period up to 12.5 days (originally specified in Release 12). The standard discontinuous receive cycle is 2.56 seconds.

Note that it is important to maximize the link budget to support IoT devices in remote rural areas. Nokia claims that the lower noise floor of narrowband LTE (200-kHz bandwidth) also known as LTE-M yields a supportable path loss of 155 dB compared to 147 dB for LTE Voice (VoLTE) and 137 dB for LTE high-speed data and video over LTE (ViLTE) (Figure 10.2).

Enhanced coverage GSM (EC-GSM) is also proposed for operators wishing to continue to support their GPRS modem user communities. EC-GSM

Table 10.1
Narrowband IoT Standards for 4G and 5G

			NB-IoT (Rel 13 Work Item)	
	NB-CIoT	**NB-LTE**	**Single Tone**	**Multitone**
Bandwidth	200 kHz	200 kHz	200 kHz	200 kHz
Downlink Multiplex	OFDMA	OFDMA	OFDMA	OFDMA
Downlink Subcarrier Spacing	3.75 kHz	15 kHz	15 kHz	15 kHz
Downlink Modulation	BPSK, QPSK	BPSK, QPSK	BPSK, QPSK	BPSK, QPSK
Uplink Multiplex	FDMA	SC-FDMA	FDMA and/or single tone SC-FDMA	Multitone, SC-FDMA
Uplink Modulation	GMSK	TPSK	Pi/2 BPSK	BPSK, QPSK, TPSK
Uplink Subcarrier Spacing	5 kHz	2.5 kHz	3.75–15 kHz	15 kHz

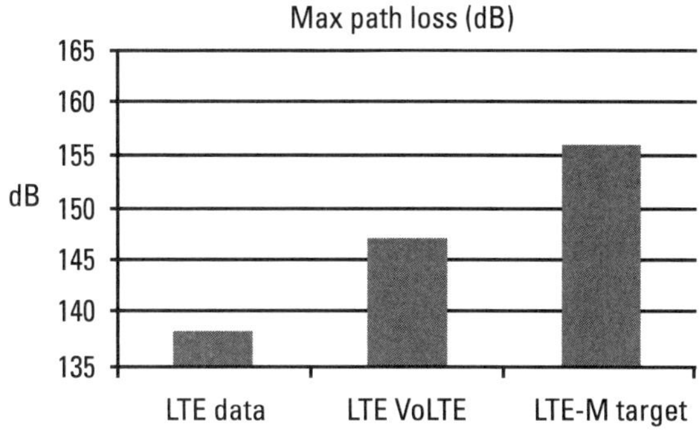

Figure 10.2 IoT power and link budgets. (With thanks to Nokia [Alcatel Lucent, now Nokia Bell Labs] [7].)

uses additional channel coding to increase the supportable path length by about 20 dB.

VHF constellations such as the Orbcomm constellation [8] can provide a cost-effective alternative to these terrestrial IoT systems. CubeSats in the sub-gigahertz ISM bands are also being used in extreme rural areas, for example, in Australia, to provide low-cost remote IoT connectivity [9].

10.10.4 Satellites and Highly Mobile Users and IoT Devices

Mobility is specified as stationary (0 km/h), pedestrian (0–10 km/h), vehicle (10 km/h–120 km/h), and high speed vehicular (120 km/h–500 km/h). All satellite constellations are effective and efficient at supporting high-mobility users without the handover overheads incurred by terrestrial networks.

10.10.5 Satellites and Large Cell Low Mobility Cells

Satellites scale from 10-km diameter cells to 2000-km cells (or more). They are therefore a cost-effective option for low mobility (and high mobility) large cells.

10.10.6 Satellites and Massive Machine-Type Communications: VHTS Flat VSATs

Perhaps not one of the more obvious applications for LEO, MEO, and GSO satellites but there are potential applications in linking factories together globally to optimize supply chain efficiency.

These applications are essentially an evolution of the traditional very small aperture terminal (VSAT) market but with the VSAT antennas replaced with flat panel active or passive arrays.

10.10.7 Satellites and Ultrareliable Low-Latency Communication

As with MMTC, not an immediately obvious win for satellites, but satellites with intersatellite switching have unique control over the end-to-end communication link. In Chapter 2, we also discussed the second-order effect of latency jitter, the amount by which latency metrics change. Although satellites introduce irreducible additional latency, they can perform better over long distances (>10,000 km) due to radio waves in free space going faster than light. Control end-to-end also means that latency parameters are all known and controllable and jitter can be closely managed.

10.10.8 Energy Efficiency and Carbon Footprint

Energy efficiency and carbon footprint targets for 5G at the time of this writing remain subject to final confirmation, but satellites have an evolving role in reducing the energy cost of backhaul and user to network and network efficiency. Reducing carbon footprint by launching a rocket in to space might not seem particularly plausible until you consider that satellites can stay in space for 20 years and have access to a free source of carbon-neutral (solar) power.

10.10.9 5G and Satellite Beam Forming

Basically 5G and satellite and automotive radar can potentially all benefit equally from innovations in antenna structures and beam forming algorithms.

10.11 Who Owns the Standards' Value?

We made the point earlier that vendors in the 4G and 5G community spend billions of dollars on research and development and cumulatively additional billions of dollars, sending engineers around the world to attend 3GPP standards meetings. This is not an altruistic process but part of an overall process to realize competitive advantage and future income from patent royalties. Historically, companies such as Ericsson, Intel, Qualcomm, Interdigital, Nokia, Cisco, and Motorola dominated the radio front end, waveform, and core network patent ownership. However, Huawei, a company owned by its employees, in common with its two network competitors, Ericsson and Nokia, spends between 12% and 14% of its enterprise income on research and development and is presently registering over 500 5G patents per year [10]. The notion that somehow China exists outside the global patent process is now well out of date, not least due to the large-scale contract manufacturing now carried out in China on behalf of more or less the whole of the rest of the world.

10.12 Satellites and Automotive Connectivity

We referenced in Chapter 6 the role that satellites will play in automotive connectivity. The automotive industry has its own standards process with which the 5G community is actively engaged [11]. The standards group includes Analog Devices, Anritsu, Audi, SAIC, BMW, Rolls Royce, Bosch, China Mobile, Continental, Daimler, Denso, Ford, Huawei, Infineon, Intel, InterDigital, Jaguar Land Rover, KDDI, Keysight Technologies, Laird, LG Telecom, muRata, Nissan, Nokia, NTT DoCoMo, Orange, Panasonic, and Proximus.

10.13 The Satellite Industry and Automotive Radar

Active engagement is also needed with the automotive radar supply chain particularly in the context of implementing E-band 5G and or very high throughput satellites where RF coexistence is precisely specified.

As referenced in Chapter 2, the 77-GHz automotive radar band consists of two subbands, 76–77 GHz for narrowband long-range radar and 77 to 81 GHz for short-range wideband radar. Compared to automotive radar at 24 GHz, 77-GHz radar provides better angular resolution due to the reduced spacing between elements. The higher carrier frequency means that the Doppler frequency increases proportionally relative to the velocity of the target, which supports higher-speed resolution. Range resolution depends on the modulated signal bandwidth; the wider the bandwidth, the better the resolution.

However, automotive radars are a safety critical component and need to be protected from RF interference. They also potentially create interference into the two E-band 5-GHz passbands below 77 GHz and above 82 GHz.

Power outputs/spectral densities for 77-GHz pulsed and frequency-modulated continuous-wave (FMCW) radar are specified by ETSI for Europe and the FCC for the United States. For the FCC, the state of the vehicle determines the restrictions on allowed output power. For a stationary vehicle, the spectral density in any direction must not exceed 0.2 μW per square centimeter in any direction. For a moving vehicle, the allowed spectral density is 60 μW per square centimeter looking forward and 30 μW per square centimeter for side-looking and rear-looking radar.

Table 10.2 shows the ETSI power output specification for automotive radar.

10.14 Satellites and 5G Data Density

Data density is not precisely described in the 3GPP standards, but various vendor estimates give an indication of expected throughput per square kilometer.

Table 10.2
ETSI Power Output Specifications for Automotive Radar

Band	76–77 GHz	
EIRP (FMCW)	50 dBm (mean)	55 dBm (max)
EIRP (Pulsed)	23.5 dBm (mean)	55 dBm (max)
3-dB Beamwidth (Typical)	5°	
Out-of-Band Emissions	73.5–76 GHz	0 (dBm/Hz)
	77–79.5 GHz	0 (dBm/Hz)

Nokia Networks, for example, suggested that today's data density of 1 Gbps per square kilometer will scale to 10 Gbps for sub-6-GHz 5G to 100 Gbps for centimeter-band 5G and 1 Tbps for millimeter-band 5G.

This provides a benchmark against which high throughput and very high throughput satellites can be measured. These higher data density applications may be better served by very high throughput constellations implemented in V-band and W-band. These density forecasts can also be used to estimate backhaul requirements.

Proposals for VHTS constellations are presently being filed with the FCC and will be a key part of satellite industry advocacy at WRC2019. A typical filing is the Boeing Constellation, rumored to be financially backed by Apple. The constellation is based on a duplex band with a lower passband (uplink) from 37.5 to 40 GHz and an upper passband at 51.4–52.5 GHz. Note that this scales down to 10-km radius cells and supports smaller spot beams due to the higher frequency/shorter wavelength.

10.15 Satellite and 5G Standards: Modulation, Coding, and Coexistence

One of the major discussion points at WRC 2019 will be protection ratios and out-of-band emissions. These are determined by spectrum band plan decisions including guard bands and allowable EIRP but also by the modulation and coding used. To meet 5G spectral efficiency targets, various higher-order modulation and coding schemes have been proposed. In practice, the physical layer is likely to be similar to LTE, although with OFDM on the uplink rather than the more heavily filtered SC FDMA used in 4G. This may mean that out-of-band emissions from user and IoT devices are higher.

Note that wider bandwidth channels will tend to have higher out-of-band emissions. A 20-MHz channel directly adjacent to a 5-MHz channel, for example, would create more interference from the wider channel into the narrower channel.

For satellite systems, the dominant requirement is to maximize RF power efficiency in space. Power efficiency is achieved by reducing the amplitude of the modulated waveform. The usual modulation used is amplitude and phase/frequency shift keying (APSK).

The combination of amplitude and phase/frequency shift keying reduces the number of power levels required to transmit a particular modulation order. This reduces the amount of linearity required in the transmit chain, trading power efficiency against a (modest) reduction in spectral efficiency.

10.16 CATs and SATs

Reading and studying standards documents is an essential requirement for telecoms engineers, particularly terrestrial telecoms engineers.

However, it is only part of the story. Performance requirements are ultimately determined not by the standards but by the devices built to conform to the standards. These often just focus on one or two of many options described in the original documents. For example, there are 32 classes of GPRS modem but only two classes are generally supported by vendors and the component supply chain.

In 4G LTE the highest category presently being sampled to smartphone vendors is a Category 18 modem from Qualcomm with a theoretical maximum data rate of 1.25 Gbps delivered over four 20-MHz aggregated LTE carriers (which few operators could potentially implement within their existing band plans). In October 2017, Intel announced their Category 19 LTE modem with a headline downlink data rate of 1.6 Mbps and plans for a sub-6-GHz 5G modem [12].

The lowest category is Cat 0. Table 10.3 summarizes the key performance parameters. Note that the 20-MHz receive bandwidth refers to the passband rather than the channel bandwidth which is typically 200 kHz.

Table 10.3
Category 0 Performance Parameters and Their Impact on Data Density

LTE Category 0	
Peak downlink rate	1 Mbps
Peak uplink rate	1 Mbps
Max. number of downlink spatial layers	1
Number of UE RF chains	1
Duplex mode	Half duplex
UE receive bandwidth	20 MHz
Maximum UE transmit power	23 dBm

The amount of data flowing across terrestrial networks in 5 and 10 years' time and the density of that data is therefore a determinant in establishing the point at which it becomes cost-effective to offload to satellite networks. This is directly analogous to the growing recognition over the past 10 years that it is often more cost-effective to offload data traffic to local Wi-Fi networks.

There are many vendor estimates of data traffic growth and, as with all estimates, these need to be approached with a measure of caution. However, it can be stated with certainty that traffic volume, traffic density and traffic characteristics, for example, the ratio of latency-sensitive to latency-insensitive traffic and traffic symmetry/asymmetry (the ratio of uplink to downlink traffic) are influenced by the mix of devices offering and receiving data to and from the network.

Table 10.4 summarizes a cross-section of modem options. Note as stated above that these are a statement of baseband capability and real-life headline data rates depend on the spectrum and band aggregation options available to the operator but it can be seen that there is a clear direction of travel in terms of offered traffic volume.

CAT 12 and Cat 16 are classed as LTE Advanced Pro and include support for LTE Licensed Assisted Access (LAA) in the 5-GHz Wi-Fi band, Citizens Broadband Radio at 3.5 GHz, for example, for the United States and Australia and Public Safety Support for the U.S. market (Band 14 in the 700-MHz band for FirstNet and Band 20 (800-MHz band in Europe and the Middle East) and Band 28 (APT 700 band in Asia).

Sierra Wireless [13] has useful additional information on LTE modem options and their technical specification.

The mix of devices supported on a network defines the volume and characteristics of the traffic offered to the network and is therefore critical to dimensioning and provisioning network bandwidth. It is also critical to dimensioning backhaul loading and determines the point at which satellite connectivity potentially becomes more economic than terrestrial connectivity.

Table 10.4
LTE Modem Categories

Speed	Cat 3	Cat 6	Cat 9	Cat 11	Cat 12	Cat 16	Cat 18	Cat 19
Downlink	100 Mbps	300 Mbps	450 Mbps	600 Mbps	600 Mbps	1 Gbps	1.2 Gbps	1.6 Gbps
Uplink	50 Mbps	50 Mbps	50 Mbps	75 Mbps	150 Mbps	250 Mbps	300 Mbps	400 Mbps
QAM	64	64	64	64	256	256	>256	>256

10.17 Satellite Backhaul for 5G

Individual devices capable of receiving more than 1 Gbps of data and capable of transmitting several hundred megabits per second will generate potentially terabytes of traffic in local backhaul.

Separate backhaul hardware is almost certainly going to be uneconomic, which implies the need to implement in-band backhauling, also described as self-backhauling.

Many links will not be line-of-sight and will suffer significant loss from wall surface absorption. The proposed use of mesh routing is only a partial solution and will absorb bandwidth and power. If satellites can deliver sufficient bandwidth at sufficiently low cost, then this is potentially a major potential source of traffic and revenue for satellite operators. It will be necessary to have almost always almost overhead coverage in order to avoid building blocking or foliage blocking.

10.18 Network Interface Standards and RF Over Fiber

There are two standards or, rather, interoperability guidance documents that describe network interfaces and network node interconnection protocols.

The Common Public Radio Interface (CPRI) [14] published by the IEEE in 2003 defines the criteria for baseband units and radio resource units (baseband and RF hardware) and theoretically at least allows distributed antenna system vendors to interface their equipment to multiple vendor products.

A parallel group of vendors produced a similar set of interconnection guidance notes a year earlier under an initiative called the Open Base Station Architecture Initiative (OBSAI) [15].

The target of 1 ms or less for physical layer latency is partly consumed by the D/A and A/D and frame delay of these interconnection nodes. Direct modulation of the RF signal on to fiber has been proposed as an alternative with a theoretic reach of 100 km (1 ms of delay) [16].

10.19 Standards and Spectrum: The HTS, VHTS, and S-VHTS Satellite Service Offer

Commentators often discuss satellite service provision as a monolithic entity but in practice satellites deliver a huge range of services across a huge range of data rates.

In this chapter, we have highlighted that device data rates, including, for example, smartphones and IoT devices produce many terabytes of data traffic.

This rapidly aggregates to exabytes as traffic is moved into the network core. Even with edge-computing architectures, traffic growth through the core is going to be directly linked to escalating data rates over the radio physical layer.

Table 10.5 summarizes how the core satellite bands could potentially aggregate together to provide sufficient bandwidth to accommodate a cross-section of bandwidth needs that are presently delivered, often expensively, over terrestrial networks. Satellite platforms could potentially scale to capture a significant percentage of this traffic, with satellites providing G/bits of connectivity from high throughput constellations in Ku-band, K-band, and Ka-band, terabits of connectivity from VHTS satellites in V-band, and potentially exhibits of connectivity from super VHTS constellations deployed in E-band.

At the risk of being contentious, we have added in 5G as a cosharing partner, an option that we explore in more detail in Chapter 12 and other incumbents including satellite TV and military radio and radar.

10.20 5G and Satellite Spectrum Cosharing

However, if the satellite industry is going to welcome 5G into its core bands in Ku-band (12 GHz), K-band (18 GHz), and Ka-band (28 GHz), and V-band and E-band, then it is only fair and reasonable for mobile broadband operators to welcome satellite into refarmed LTE spectrum from 450 MHz to 3.8 GHz. Table 10.6 lists the 14 super bands, which 5G could (and we argue should) coshare.

Effectively, we are saying that the satellite industry has a physical layer that sits comfortably within existing 4G and 5G channels and passbands and the direct implication of this is that band sharing is possible from 450 MHz to 95 GHz.

The table extends from UHF (LTE/5G Band 31) to E band (72–77 GHz, 81–86 GHz, and 92–95 GHz) and scales from a 5-MHz passband at 450 MHz to a 5-GHz passband in E-band. It could potentially scale down to include

Table 10.5
HTS, VHTS, and SVHTS Satellite Services for the 5G Industry

HTS			VHTS		SVHTS	
Ku-Band	K-Band	Ka-Band	V-Band		E-Band	
12 GHz	18 GHz	28 GHz	37.5–40 GHz	51.4–52.5 GHz	72–77 GHz	81–86 GHz
SAT TV	SAT TV					
Military radio						
5G?	5G?	5G?	5G?	5G?	5G?	5G?

Table 10.6
5G and Satellite Band Sharing from UHF (or VHF) to E-Band

5G+ Satellite Refarm					**L-Band, S-Band, C-Band Refarm**			**Satellite + 5G + 5G In-Band Backhaul Cosharing**					
450 MHz	600	700	800	900	L-band	S-band	C-band	Extended C	Ku-band	K-band	Ka-band	V-band	E-band
Band 31	Existing LTE/5G core bands, band share with satellite?				Existing LTE/5G core bands, band share with satellite?			Existing and future satellite +5G band share					

VHF with the Orbcomm 1 + 1-MHz passbands refarmed as five 200-kHz 5G NB IoT channels.

10.21 Implications of 5G and Satellite Band-Sharing on Regulatory and Competition Policy

This would imply a reasonably radical repurposing of all existing 4G and 5G and satellite spectrum and implies a major shift of focus in the present standards and spectrum allocation and auction process.

10.22 Physical Layer Compatibility

Our suggestion would be that there could and should be an S-LTE and S-5G standard that implements an APSK-based physical layer that can comfortably coexist within all existing sub-3.8-GHz LTE and 5G passbands. This should be relatively easy as APSK requires less linearity (lower out-of-band emissions) than the multiplexed OFDM QPSK used in 4G and 5G. Conversely, above 3.8 GHz, consideration needs to be given to implementing a power efficient rather than spectrally efficient physical layer for the direct to Earth direct to space uplink and downlink using APSK rather than QPSK.

Note that different physical layers already cochare spectrum in 4G and 5G, for example, NB-IoT 200-kHz channels within 5-MHz LTE. The guiding principle is that adding in additional physical layer support must have no impact on protection ratios and/or device radio frequency front-end components. In practice, APSK will pass happily through any existing and potential future RF front-end switch path, filter path, and RF power amplifier LNA.

This makes it completely possible to produce smartphones that can look upwards to the sky and connect with LEO, MEO, and GSO satellites. There are then two connectivity modes to standardize depending on whether the user and or IoT device have passive (low-cost) flat VSAT antennas or active flat VSATs.

10.23 Passive Flat VSAT Standards

Passive flat VSAT antennas look directly upwards through a narrow cone of visibility. At higher latitudes, there will be a regular handover between LEO and MEO satellites passing overhead. High-count LEO and MEO constellations would provide effectively continuous connectivity (the user or IoT device would not detect that a handover had taken place). Low-count constellations would provide periodic rather than continuous connectivity; the Myriota CubeSats traversing the Australian outback every 90 minutes to collect IoT data provides a contemporary example. At the equator, there will be continuous connectivity from directly overhead geostationary satellites. In all cases, all unwanted signal energy arriving at an acute angle outside the vertical cone of visibility is invisible to the receiver (and transmissions go directly upwards).

10.24 Active Flat VSAT Standards

Active Flat VSAT antennas scan from horizon, choose the optimum LEO, MEO, or GSO connection and actively null out unwanted signal energy. At the equator, best connect would probably be from GSO constellations particularly a directly overhead GSO. At higher latitudes, the best connect would probably be from a directly overhead or nearly overhead LEO or MEO, although if heavy rain fade hits the vertical link budget, the active flat VSAT would look at lower elevation angles.

10.25 In-Band 5G Backhaul and Satellite

Additional standards work items could and should include in-band 5G backhaul.

10.26 ESIM and BSIM Standards: Model T Connectivity

There could and should be a work group that produces a specification for a base station in motion (BSIM). This is effectively an extension of the existing Earth stations in motion (ESIM) work stream but specifying satellite connectivity to cars, trucks, trains, and planes integrated with terrestrial 4G and 5G. This is what we call Model T connectivity, the principle of shipping 6.6 million Ford motor cars, trucks, ambulances, fire engines, police cars, garbage trucks, and tanks with satellite and 5G connectivity.

10.27 Specifying Network Power Efficiency and Carbon Footprints

Satellites connectivity minimizes the signaling overhead associated with supporting mobile and highly mobile users and IoT devices.

Passive and active flat VSATs also remove the need for power control. Counter intuitively this would make terrestrial and satellite networks more power efficient because power control overheads are avoided, and power amplifiers can be run at their optimum operating point. The allocation of power to individual users and IoT devices is realized in the time domain.

10.28 CATSAT Smartphone and Wearable SAT Standards: Tencent Telefonica and Other Unexpected Outcomes

Finally, all the above couple into user and IoT device standards. For example, today there would be CATSAT O, CATSAT 1, and CATSAT 18 modems supported by RF front ends that are transparent to the LTE, 5G, and satellite physical layer.

This would mean that users and IoT devices would always be connected anywhere in the world. EBITDA and enterprise value in the satellite industry and mobile broadband industry would be an order of magnitude higher than it is today. Imagine the headlines, Inmarsat buys Google, Intelsat buys Facebook, OneWeb merges with Alibaba (One Baba), Telefonica and Tencent merge to form Tencent Telefonica: a brave new world and all eminently deliverable.

10.29 Summary

Thousands of engineers spend thousands of hours discussing and writing mobile broadband and local area and personal area standards. The Bluetooth 5.0 specification is 2,800 pages long and is one of the simpler standards documents. The satellite industry has nothing on a similar scale. Standards and scale together deliver cost and performance and interoperability benefits.

The 5G standards process started with Release 15 and is continuing with Releases 16 and 17 in parallel with LTE Advanced. The 5G standard as discussed in other parts of this book (and our earlier book) is more finely resolved in the time domain (0.1-ms mini frames). This reduces over the air latency and theoretically at least improves power efficiency. In the frequency domain, a flexible OFDM subcarrier structure scales from 15-kHz subcarriers to 30, 60, 120, 240, and 480-kHz subcarriers to allow scaling to channel bandwidths of the order of 250 MHz, 400 MHz, or in the longer term, 1 GHz.

In the phase domain, the physical layer scales to 1,024 QAM to help replicate a fiber-like experience for fixed wireless and mobile wireless users and devices and to support the headline target per use per device rates of 10 Gbps.

It is unlikely that satellites will have the link budget and flux density to deliver these data rates to smartphones optimized to receive signal energy in the horizontal rather than vertical plane. It is entirely possible for satellites to deliver these data rates and potentially higher data rates to active or passive flat panel arrays optimized for vertical coverage, for example, active and passive flat VSAT conformal antennas built into car, truck, train, or plane roofs and into smartphones and wearable devices. This allows the satellite industry to scale into volume consumer markets and to become a crucial art of the added value of next-generation smartphones and wearable devices.

5G and satellite standards also need to coexist with legacy fixed wireless standards implemented into C-band and with Wi-Fi in the 2.4-GHz, 5-GHz, and 60-GHz bands, defined within the IEEE standards process and with Bluetooth Low Energy 5.0, defined by the Bluetooth Special Interest Group.

Guided media standards continue to move forward, and user experience expectations will continue to be determined by fixed connectivity performance. At the other extreme, proximity technologies (within 5 cm) such as Near-Field Communication (NFC) [17] continue to evolve and are essential to many pairing and transaction applications.

Within all terrestrial networks, standards are a major determining factor in spectrum allocation and the setting of intersystem and intrasystem protection ratios. However, ultimately the mix of devices supported in the network shape the offered traffic and hence the spectral properties of the occupied bandwidth.

Satellites can help the 5G industry deliver on many of the objectives specified in the standards and related use cases. This includes meeting energy efficiency and carbon footprint targets but also delivering rural coverage and IoT connectivity and critically, adding additional value to smartphones and wearable devices. This should motivate standards engagement between the 5G and satellite community.

Ultimately, the purpose of standards is to create an ecosystem in which operators and their supply chain make sufficient profit to sustain research and development and manufacturing investment and provide an adequate return to shareholders. This does not always happen, which brings us to the topic of our next chapter.

References

[1] https://www.qualcomm.com/news/onq/2017/08/02/top-5-drawbacks-contribution-counting-3gpp-dont-count-it.

[2] http://tec.gov.in/pdf/Studypaper/S_UMTS_Final.pdf.

[3] http://www.ic.gc.ca/eic/site/smt-gst.nsf/eng/h_sf09857.html.

[4] http://www.rttonline.com/tt/TT2010_011.pdf.

[5] 3GPP TSG RAN meeting#75, RP-170132.

[6] https://www.qualcomm.com/invention/technologies/lte/multefire

[7] https://resources.ext.nokia.com/asset/200175.

[8] https://www.orbcomm.com/

[9] http://myriota.com/.

[10] http://5gaa.org/.

[11] http://www.theregister.co.uk/2015/03/05/huawei_to_build_5g_patent_book/.

[12] https://newsroom.intel.com/news/intel-introduces-portfolio-new-commercial-5g-new-radio-modem-family/.

[13] www.sierrawireless.com

[14] http://www.cpri.info/.

[15] http://www.obsai.com/specs/OBSAI_System_Spec_V2.0.pdf.

[16] http://www.apichip.com/.

[17] https://nfc-forum.org/what-is-nfc/.

11

U.S. Bankruptcy Procedure

11.1 A Financial Overview of the Telecom Industry and Its Associated Supply Chain

In this chapter, we look at the numbers underlying the telecommunications industry, its satellite subset, and those of the various associated supply chains.

In previous chapters, we suggested that adding satellite connectivity to smartphones and wearable devices was achievable and would have a transformative impact on mobile broadband operator and satellite operator EBITDA and enterprise value. Passive and active Flat VSATs were proposed as the key technology enabler of this transformation process.

The transformation depends on the mobile broadband industry and satellite industry cosharing spectrum from very high frequency (VHF) to E-band. Understandably, this is a radical proposition, but we argue that the financial dynamics of the telecommunications industry require a different approach to how we use and value and share spectrum.

11.2 Lessons to Be Learned from Past Financial Failures: Chapter 11 as a Revolving Door

There have been some notable financial failures in the satellite industry over the past 20 years. In January 2002, Iridium filed for bankruptcy protection under U.S. Chapter 11 [1] after defaulting on $1.5 billion of loans. It was bought by private investors for $25 million. Iridium had failed because the original market plan had failed to factor in the rapid growth of low-cost GSM networks and devices in the 1990s.

The company emerged from Chapter 11 and is now a profitable and respected part of the satellite communications industry as can be seen from Figure 11.1. At the time of this writing, the company had launched 30 replacement satellites into low Earth orbit (LEO), one of the fastest constellation replacements ever in the history of the industry.

In February 2002, Globalstar followed Iridium into Chapter 11 with liabilities of $3.4 billion. The company also refinanced though its recovery has been hampered by RF hardware failures.

In parallel, Teledesic announced that their constellation development would cease. Backed by Craig McCaw [2] and Bill Gates, the original plan announced in 1994 was for a constellation of 288 LEO satellites and the first satellite was launched in 1998. The system was promoted as offering fiber optic like links to customers around the world. The constellation would be using Ka-band.

In 2012, Light Squared filed for bankruptcy. The company had planned a hybrid LTE terrestrial and satellite constellation in L-band but struggled to accommodate the GPS industry, which considered that the constellation would compromise GPS receiver performance. This dispute cost Light Squared $1.8 billion. The company remerged from Chapter 11 in February 2015 under the control of its biggest lender, Dish Network Corporation, and has subsequently come under the control of Harbinger Capital Partners. The company has been renamed Ligado, the Spanish for connected, and hopes to target Latin American markets and underserved U.S. markets, although, at the time of this writing, a threat of litigation had surfaced from companies operating geostationary environmental monitoring satellites.

The two lessons that can be learned from these four examples is that competing terrestrial service offers can scale quickly and achieve cost floors far lower than those achievable from satellite systems. Additionally, interference disputes can derail even apparently robust business models.

Scale is also essential if consumer products are important to the product offer. At the component level, failure to be listed as a vendor in a next-generation Samsung or Apple smartphone can result in a huge and occasionally catastrophic loss of share value. The semiconductor component supply chain, a sector with a turnover of $400 billion in 2017, is understandably wary of diverting research and development and production resources away from these key customers.

With these cautionary tales ringing in our ears, it is time to look more closely at the financial dynamics of the 5G and satellite industry and other stakeholders.

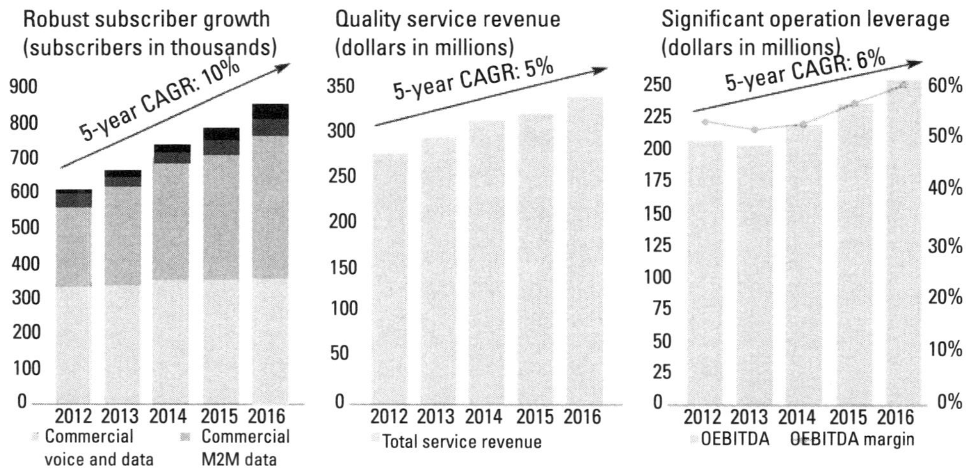

Figure 11.1 Iridium growth and revenue.

11.3 The Size of the Telecoms Industry

The global telecommunications industry generated revenues of about $2.15 trillion in 2016. This represented a marginal increase on the 2015 number of $2.11 trillion. The total EBITDA exceeded $700 billion, while aggregate capital expenses (CAPEX) was just above $600 billion. The industry's combined enterprise value was of the order of $4.6 trillion, of which $1.6 trillion was debt. The average debt to equity ratio was just above 100%, while return on equity stood at a healthy 12.2%. Although customer growth has stalled and even gone into reverse in some markets, suggestions that the industry is struggling appear to be unfounded.

This number, $2.15 trillion, includes pay TV and infrastructure services as well as both fixed and mobile communications. The vast majority, about $1.75 trillion, comes from traditional telcos such as AT&T, Deutsche Telekom, and Vodafone, with most of the balance being attributable to pay TV companies such as Comcast, Liberty Global, and Sky. The remainder arises from many ancillary activities, such as wholesale carriage and infrastructure services, including satellite connectivity. The aggregated numbers and ratios in the tables are based on representative samples of numbers reported by publicly quoted members of each subset of the industry. In the interest of simplicity, the telco group shown here consists of just the 10 largest operators, all of whom have annual revenues of over $50 billion. For the record, these are AT&T, Verizon, China Mobile, NTT, Vodafone, Deutsche Telekom, Softbank, China Telecom, America Movil, and Telefonica. Together, they account for approximately half the global industry.

11.4 The SATS and Other Entities

The satellite group we show here consists of nine separate entities. Broadly speaking, these can be further divided into two subgroups, geostationary operators and LEO companies. Echostar, Eutelsat, Inmarsat, Intelsat, SES, and ViaSat belong to the first set; Globalstar, Iridium, and Orbcomm belong to the second. For the moment, we can make no sensible comments any of the new LEOs, such as OneWeb, Space X, or LeoSat, as all are pre-revenue. The internet/OTT companies we have included are the GAFA group, Google (Alphabet), Amazon, Facebook, and Apple, although the reader should note that other businesses such as Alibaba and Tencent are comparable to some of these in scale, if not in reach.

In Table 11.1, we have shown two versions of the telco group, the second of which excludes China Mobile. We have done this because its $64 billion cash balance makes it entirely atypical: the nine remaining companies have, on aver-

Table 11.1
Financial Comparisons

Billions of U.S. Dollars	Telecom Group	Telecom Group*	GAFA Group	Satellite Group
Revenue	901.84	789.80	276.05	12.71
EBITDA	274.16	233.86	82.68	6.75
Net income	73.59	56.16	29.78	2.52
Enterprise value	1,492	1,378	1,722	55.2
Shareholders' equity	607.5	463.5	349.9	13.49
Net debt	508.4	572.6	−157.4	24.72
Capital expenditure	95.57	82.92	22.57	1.27
Debt to equity	83.7%	123.6%	−45.0%	183.6%
Return on equity	12.1%	12.1%	8.5%	7.7%
EBITDA margin	30.4%	29.6%	30.0%	55.7%
Capital intensity	10.6%	10.5%	8.2%	39.6%
EV/EBITDA	5.44	5.89	20.8	7.7
EV/revenue	1.65	1.75	6.2	4.4

*Eliminating China Mobile.

age, over $60 billion of net debt. Each of the three groups has distinct financial characteristics.

Perhaps the most obvious difference between the groups is the scale of these businesses. On average, each of our 10 large telcos has annual revenues of $90 billion, EBITDA of $27 billion, debt of $50 billion, and a valuation of about $150 billion. The numbers for the GAFA group are $69 billion, cash of $39 billion, and an enterprise value of $430 billion. The satellite companies are dwarfed by these numbers, with average revenues of $1.4 billion, EBITDA of $750 million, debt of $2.75 billion, and an enterprise value of $6.1 billion.

11.5 The Satellite Supply Chain

The dissimilarity in financial dynamics can also be seen when considering the industry's supply chain. Apart from the satellite group, which remains unchanged, we have used smaller samples from each industry group, confident that our selections are representative. The U.S. Aerospace group consists of Boeing, Lockheed Martin, and Northrop Grumman; the European Aerospace companies are BAE Systems, Airbus Industries, and Thales; the vendor group consists of Ericsson, Huawei, and Nokia; and the automotive manufacturers are Ford, General Motors, and Toyota. Table 11.2 shows the financial highlights of these groups. In all cases, the numbers used relate to the 2016 financial year.

Table 11.2
Other Stakeholders

Billions of U.S. Dollars	Satellite Group	U.S. Aerospace	European Aerospace	Network Vendors*	Automotive Group
Revenue	12.71	166.4	104.2	124.54	537.71
Operating profit	3.95	14.58	5.83	9.35	38.70
Net income	2.52	12.40	3.10	4.80	30.83
Enterprise value	55.2	239.4	132.1	117.9	632.7
Shareholders' equity	13.49	7.68	13.01	34.86	241.6
Net debt	24.72	26.91	39.94	–1.03	297.97
Capital expenditure	1.27	3.60	4.18	5.53	27.52
Operating margin	20.96%	8.76%	5.60%	7.51%	6.75%
Return on equity	18.75%	161.42%	23.82%	13.77%	12.76%
Debt to equity	183.6%	350.4%	307.0%	–2.96%	123.32%
Capital intensity	32.60%	2.16%	4.01%	4.46%	4.80%
Private venture (PV) research and development /sales	Not meaningful (NM)	3.80%	5.28%	14.68%	4.36%
Revenue per employee (U.S. dollars in thousands)	960	529	371	316	726

*As Huawei is privately owned, its EV has been estimated based on the Nokia and Ericsson numbers.

11.6 Financial Comparisons

There are a lot of numbers here and it is perhaps not easy to grasp the variations in the various industry groups. Figure 11.2 shows the sizes of the average constituent of each industry group measured by revenue, debt, shareholders' funds, and capital expenditure.

Figure 11.3 highlights the second main point of difference. Broadly speaking, the capital structures are entirely dissimilar. The telcos are indebted, but none seems to be excessively burdened by debt: EBITDA at the two largest net borrowers (AT&T and Verizon) covers their annual interest bill some 10.3 and 9.8 times, respectively, while even Telefonica and Softbank manage 6.8 times and 5.8 times, respectively. By contrast, Intelsat struggles and fails to cover its charge more than twice.

It should be noted that Figure 11.3, debt to equity, is less helpful than it might be, due to the impact on the scale of the U.S. aerospace industry. At 350% it reduces all other ratios to bit part players. That figure of 350% is anomalous as it stems from some aggressive financial engineering at Boeing, which, by reducing shareholders' funds from $6.3 billion to $817 million last year, has raised its return on equity from 82% to 600%. This, in turn, suggests that caution ought to be exercised when considering these figures.

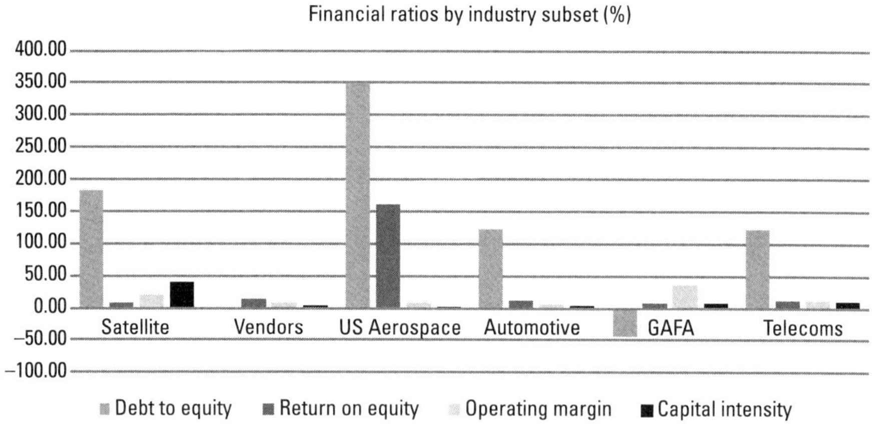

Figure 11.2 Financial metrics by industry subset.

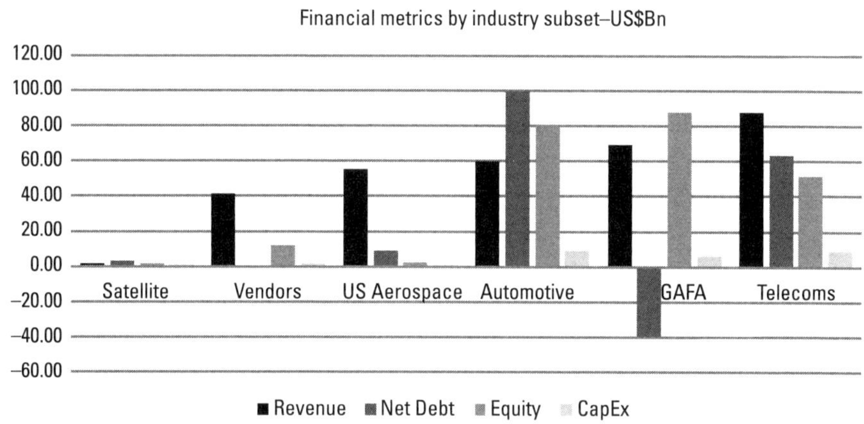

Figure 11.3 Financial ratios by industry subset percentage.

Despite the graphics, there are so many striking comparisons in Figures 11.2 and 11.3 that it is hard to know where to begin. Apart from the obvious discrepancy in size, there are extreme comparisons between these groups, most especially in the areas of profitability and capital structures.

11.7 The GAFASATs and Automotive Majors

At the same time, there are other points to note. Entities like Facebook and Google derive revenues almost minute to minute, while the aerospace giants plan their affairs decade by decade. Between these extremes lie the telcos with their monthly billing cycles, then the automotive and vendor communities, and finally the satellite sector, with their focus on long-term recurring revenues. These variations stem from the differing size of each group's target market. Facebook and Google measure their customers in billions, large fractions of the world's population, while the telecom majors have anything from 50 to 500 million customers. One order of magnitude lower, we find the automotive companies. Another four or five orders further down, we get the telecom vendors, which serve perhaps 500 companies, while some aerospace companies measure their customers on the fingers of two hands. These differences are reflected in valuations, and it should be noted that the new generation LEOs are more likely to share the characteristics of a telco than those of a GSO.

Taking a closer look and starting with the telco supply chain, it is clear that these are huge businesses, dwarfing the satellite group and, at least in terms of annual turnover, being roughly on a par with the companies that they supply. The telecom supply chain continues to experience rapid change, driven by technology and the shape and structure of its customer base. (Gone are the days of monopolies, except of course in the geostationary sphere.) It is fair to say that today there is no one company that is typical of the industry. In the 6-year period that we chose for our trends, Nokia sold off its handset business and acquired Alcatel, which was once, in the years following its merger with Lucent, the clear market leader. Ericsson underwent less a dramatic metamorphosis, but its shift from hardware to services was indicative of a sea change.

11.8 The Huawei Factor

The name of that sea change is Huawei. The Chinese business has more than doubled in size over the last 6 years from $32 billion to $75 billion with profits and cash balances keeping pace. The increases are even more impressive in local currency, with revenues up from 204 billion to 522 billion in Chinese yen, operating profits up from 18 billion to 48 billion in Chinese yen. Bitter competitors complain of reverse engineering and plagiarism, but Huawei spends an industry average 15% of sales on PV research and development and its advances seem genuine enough.

There are several thousand independent telecom operators in the world at present, but no more than 100 with any great market presence and only really 20 that absolutely matter, from the perspective of a telecom supplier; get the 10 that we mentioned above as customers, and then add some others like Orange, BT, Telecom Italia, and KDDI. For the defense contractor, the task of marketing is even simpler.

11.9 The Defense Sector Supply Chain

Lockheed is a case in point: more than 70% of its annual output goes straight to a single customer, the U.S. Department of Defense. Indeed, at the time of this writing, one single program, the F-35 Lightning II, accounts for almost one-third of that total. According to Lockheed's 2015 10K filing, "the F-35 is designed to be an affordable, superior multirole stealth aircraft." With an estimated fully operational price of $251 million per copy (U.S. DoD, 2015 estimate), one might wonder what an unaffordable aircraft might cost. However, the fact is that these programs seem to go on forever: Lockheed still highlights the contribution to revenues from the F-16, an aircraft that first saw service in 1973 and is still being revamped today. We can think of very few other types of enterprise that enjoy such a steady stream of predictable, recurring revenues.

Although mainly an aircraft specialist, approximately one-fifth of Lockheed's revenues come from its Space Systems division. Over the last 5 years, this has generated average revenues of $9.07 billion and operating profits of $1.18 billion. Again, the emphasis here is on military work, with Trident II, the U.S. Air Force's space-based infrared system, and GPS III being notable.

Boeing's main focus is on commercial aircraft (between 60% and 70% of revenues and a slightly higher percentage of profits over the last 5 years), but it too has a sizeable presence in space. Its Network and Space Systems division is a $7 billion to $8 billion business, with an involvement in military programs such as GPS III and the Wideband Global SATCOM constellation (12 LEOs). In addition, it has several important commercial programs for customers including MexSat, SES, and ViaSat.

The European Aerospace Group is directly comparable to its U.S. peers. For Lockheed, insert BAE, which has a similar 70% of all revenues arising from government customers. In this case, the customers are rather more diverse: the governments of the United Kingdom, United States, and Saudi Arabia and the Eurofighter Consortium. Airbus is analogous to Boeing: mainly civil, but with some satellite and military programs thrown in. It is interesting to note that Airbus' civil backlog now stands at over €1 trillion ($1.24 trillion), equivalent to more than 15 years' revenues, at the current run rate of $82 billion.

Can these behemoths realistically be expected to achieve the flexibility and responsiveness needed to service demanding customers such as the telco group, the automotive industry, or any group of customers where individuals' remunerations are directly tied to performance? These are interesting questions, and from my experience of analyzing both defense contractors and telecom operators, it is not clear that they can be answered in the affirmative.

11.10 The Satellite Supply Chain

The satellite group is hugely capital intensive, and consequently, the ratio of capital expenditure to revenues (capital intensity) is more than six times higher than that of any other group. Similarly, average revenue per employee is nearly twice that of the second most efficient group (U.S. aerospace). This is because the satellite operators have only a limited number of customers and rely on external contractors to manufacture, launch, and sometimes operate their satellite fleet, rather than performing these functions in house. SES is the most extreme example within the group; at the end of 2016, it had just 69 employees, which gave it a figure of $33 million in revenue per employee.

Mention of this outlying statistic highlights another aspect of the satellite group; its nine constituents are hardly homogenous. Table 11.3 highlights key numbers and ratios, with the GSOs appearing first, followed by the LEOs and the group aggregates. Looking at these numbers, it becomes clear that the industry's reputation for excessive borrowings is almost entirely due to Intelsat, or rather, the private equity investors who landed the business with such an unhelpful capital structure. Intelsat owes more than half the industry's total debt and were we to strip it out; the GSO subgroup's debt to equity ratio falls to below 80%.

Table 11.3
Satellite Industry Financial Comparisons

Millions in U.S. Dollars	Revenue	EBITDA	CAPEX	Net Debt	Enterprise Value	Debt to Equity	Revenue/ Employee
GSO subset							
EchoStar	3,057	859	722	567	5,418	15.44%	0.764
Eutelsat	1,619	1,252	364	4,171	8,681	144.67%	1.755
Inmarsat	1,329	795	150	2,290	5,667	160.65%	0.762
Intelsat	2,188	1,616	1,980	13,532	13,853	–2,982.42%	1.903
SES	2,284	1,664	1,602	1,265	12,673	47.29%	33.107
ViaSat	1,515	285	485	762	4,269	93.78%	0.352
	11,992	6,471	5,303	22,587	50,561	205.06%	0.984
LEO subset							
Globalstar	95	15	7	762	1,767	322.50%	0.277
Iridium	434	226	406	1,247	2,215	81.37%	1.777
Orbcomm	187	41	28	124	710	40.32%	0.404
	716	282	441	2,133	4,692	102.73%	0.682
Combined total	12,708	6,751	5,744	24,717	55,254	187.21%	0.960

11.10 The LEOs

The LEO operators have a higher debt level than that 80% number, due for the most part to the debt taken on by Globalstar after its flirtation with bankruptcy. A net debt to EBITDA ratio of over 40 is clearly not healthy, but this is not a major business and even were it to disappear back into Chapter 11, that would hardly jeopardize the overall state of the industry. (For the record, the $762 million net debt seen for both Globalstar and ViaSat is a coincidence, not a typo.)

Every single company that we have covered in this brief overview has revenues that are greater than the satellite group taken as a whole. Only Intelsat, with its debt mountain, looks in any way formidable to the denizens of the city's mergers and acquisitions (M&A) departments, and even then, it would only occasion mild indigestion for most would-be predators. Most commentators, when considering the vulnerability of this group to unwanted advances, have assumed that the likely predators will come from the telecom group. More than a few telcos already have some involvement in the satellite market and some, in the past, were shareholders in businesses like Intelsat and Inmarsat. (Those with long memories may recall that Vodafone was once a shareholder in Globalstar.)

This assumption that telcos are the most likely buyers seems reasonable, but there may be other interested parties who are better positioned. Certain members of the GAFA group have expressed an interest in using satellites to bypass traditional communications networks, most notably Facebook. This makes absolute sense, given that the one obvious weakness in their business models is that they require uninhibited access to someone else's networks to offer their services. For the record, Apple has enough cash to pay a 25% bid premium over the group's $28 billion valuation and still buy the whole lot, twice over. Alphabet, Google's parent, could do that too and still have enough financial firepower to acquire most of the world's satellite TV companies. Facebook may not have such deep pockets as these other two; net cash was a mere $8.9 billion at the end of December 2016, but with its stock rated at more than 40 times earnings, it has the potential to mount a realistic bid for any or all the nine.

Several of these entities are sitting on spectrum assets that they are not exploiting as aggressively as they might and that, in turn, suggests that their independence may be in doubt. This could have implications to suppliers and customers alike.

11.10 Summary

The mobile operator community and satellite operators (excluding Intelsat) are fully geared, although not overgeared. However, they are dwarfed by the potential financial firepower of the GAFA quartet and Alibaba and Tencent. The satellite industry is two orders of magnitude smaller than the mobile broadband industry and automotive industry. This may change once the NEWLEOs launch and start building their customer base, but this depends on creating business models that foster cooperation rather than commercial conflict.

The mobile broadband industry and the satellite industry could transform the economics of mobile and fixed broadband delivery and the delivery economics of the global IoT market by cosharing spectrum, but will they do it?

References

[1] http://bankruptcy.findlaw.com/chapter-13/chapter-11-bankruptcy.html.

[2] http://www.bornrich.com/craig-mccaw.html.

12

Mutual Interest Models

12.1 Introduction

In this chapter, we summarize the tension points and touch points between the 5G industry and satellite industry and their mutual potentially positive interaction with other stakeholders including the automotive industry and Web-scale companies such as Google (Alphabet Group), Apple, Facebook, Amazon, Alibaba, and Tencent.

We review the technical and commercial common interest between these industries and quantify the financial benefits of a collaborative approach to sharing assets including spectral assets, space and terrestrial assets and customer assets.

In Chapter 9, we identified problems that the 5G industry needs to solve and how the satellite industry can help. Conversely, we described the problems that the satellite community needs to solve and how the 5G industry and other stakeholders can help. We stressed that the satellite industry needs scale. The 5G industry needs to reduce the cost of delivery, provide coverage to existing not spots, improve power efficiency, and meet carbon footprint targets.

Note that we are including Wi-Fi and Bluetooth and NFC as an integral part of 5G but are also arguing the case for adding LEO, MEO, and GSO coverage to the 5G delivery mix.

12.2 Spectrum Touch Points and Tension Points

Table 12.1 shows the spectrum touch points and tension points from very high frequency (VHF) to V-band and E-band. Changes of use and/or change to access rights to spectrum will always involve one set of users having to accept another set of users either joining them or replacing them. This will usually be contentious and will often trigger legal and commercial disputes.

Table 12.1(a)

VHF to E-Band Touch Points and Tension Points

Touch Points and Tension Points Sub 3 GHz						
UHF Sub 1 GHz			**L-Band**	**S-Band**		
450 MHz	600 700 800	900	1.6 GHz	2 GHz	2.3 GHz	2.7 GHz
Band 31 452-457	Band 71 600 MHz		Band 21 1447-1462 1495-1510	Band 1 LTE FDD	Band 30 LTE FDD 2305-2315 2350-2360	Band 7 LTE FDD LTE TDD
			Iridium Globalstar Ligado SAS	Inmarsat EAN	Globalstar	
5G Refarm	5G Refarm	5G Refarm	5G Refarm AWS 1.8 GHz LTE 1.9 GHz LTE	5G Refarm	5G Refarm	5G Refarm
PMSE		ISM 902-908	GPS		2.4 GHz Wi-Fi	
						Long-range weather radar 300 km Heavy rain 2.7-2.9 GHz
Channel Raster and Passband						
25 KHz PMSE						
5 MHz LTE	10 MHz LTE	10 MHz LTE	10 MHz LTE	20 MHz LTE	10 MHz LTE	20 MHz LTE
5 MHz	10-45 MHz	35 MHz	75 MHz	75 MHz	?	75 MHz

Table 12.1(b)

VHF to E-Band Touch Points and Tension Points

Touch Points and Tension Points >3 GHz											
C-Band		**X-Band**	**Ku-Band**	**K-Band**	**Ka-Band**	**V-Band**		**E-Band**			
			High Throughput Satellites (HTS)					**Very High Throughput Satellite (VHTS)**			
3.4 GHz 3.8	5.6 GHz	8-12 GHz	12 GHz	18 GHz	28 GHz	50 GHz	60 GHz	71-76 GHz	77-81 GHz	82-87 GHz	92-95 GHz
Band 22 FDD Bands 42,43,48 TDD	?		GSO MEO LEO	Feeder links	GSO MEO LEO	GSO MEO LEO					
Satellite TV 3.7-4.2 GHz		Military radio and radar	Military radio and radar SAT TV	Military radio and radar SAT TV 18.13 18.8 19.7 20.2	P to P Backhaul				Short Medium Long Range Auto Radar		
5G Refarm FDD/TDD	5G New FDD/TDD		5G New FDD	5G New FDD	5G New FDD	5G New FDD	5G New TDD	5G New FDD		5G New FDD	5G New TDD
	5 GHz Wi-Fi + 802.11p										
	Medium Range Weather Radar Light Rain 5250 5725 MHz	Short Range Weather Radar Light Snow 9300 9500 MHz									
Channel Raster and Passband											
20 MHz	100 MHz		250 MHz		250 MHz	500 MHz	1 GHz	1 GHz	1 GHz	1 GHz	1+2 GHz
400 MHz	?		2 GHz	2 GHz	3.5 GHz	3.5	3.5	5		5	5

Note 5G FDD bands align with satellite FDD bands. 5G TDD bands align with 5 GHz and 60 GHz Wi-Fi and the 92-95 GHz band

Examples shown in Table 12.1 include two-way radio in the VHF and ultrahigh frequency (UHF) band, IoT users in the subgigahertz ISM bands, terrestrial TV between 450 and 800 MHz, heavy rain long-distance radar at 2.7 GHz, satellite TV between 3.7 and 4.2 GHz, light rain medium range radar at 5 GHz, military radar, military radio, and high-definition satellite TV at 12 GHz, superhigh and ultrahigh definition satellite TV at 18 GHz, point-to-point and point-to-multipoint backhaul at 28 GHz, and short-range, medium-range, and long-range automotive radar at 77 to 81 GHz.

Assuming that 5G delivers higher peak data rates, more capacity, lower-cost, and better power efficiency than LTE (admittedly an assumption), then it could be expected that over the next 4 to 10 years, all bands between 450 MHz and 3.8 GHz (existing LTE) will be replaced by 5G. This is called refarming.

As OFDM is used in 5G user devices (replacing the more heavily filtered SC FDMA used in LTE), then it could be anticipated that out-of-band emissions could increase. Wider bandwidth channels, for example, 20-MHz LTE or 20-MHz 5G will also increase interband, intersystem interference in these lower bands.

At the lower end of the table, we show how channel bandwidths and passbands increase in size as frequencies increase. At 450 MHz, a 5-MHz LTE/5G channel sits within a 5-MHz passband, this increases to 5-MHz and 10-MHz channels implemented in a 45-MHz passband (APT bands at 700 MHz), in L-band, and 10-MHz or 20-MHz LTE/5G sit within 60-MHz or 75-MHz passbands with similar bandwidths and passbands in S-band (Band 1 at 2.1 GHz and Band 7 at 2.6 GHz). In C-band, this increases to a 400-MHz passband (3.4 to 3.8 GHz), although regional and country-specific differences in TDD/FDD band plans frustrate potential global scale economy. There are also country-specific and region-specific coexistence issues, for example, coexistence with citizens broadband radio in the United States, fixed wireless (using Wi-Max TDD), for example, in Australia and satellite TV in many markets. The higher end of C-band includes 5-GHz Wi-Fi with 802.11p for automotive V2V (vehicle to vehicle) and V2X (vehicle to network [V2N]) and then a cross-section of military radio and radar systems through to the lower end of K-band. Note that each new generation of radar system generally has higher transmit power than the previous generation and a wider receive passband and improved receive sensitivity. Together, these improve the range, accuracy, and resolution of the radar but increase out-of-band emissions and make the radar more susceptible to adjacent channel and adjacent band interference.

The discussion points for the 12-GHz band, apart from coexistence with satellite TV and military radio and radar, is whether the new high-count LEO constellations, specifically OneWeb, can coshare the same spectrum with MEO and GSO constellations (and satellite TV and military radio and radar). The cosharing is predicated on the ability of progressive pitch angular power separation combined with power control and handover to prevent unwanted signal energy getting in to the other radio systems cosharing the band. This has been challenged by some existing incumbents (see, for example, the Asia Broadcast Satellite Case study in Chapter 7). The same discussion is ongoing for K-band, for example, for feeder links at 18 GHz and for Ka-band. Part of the motivation for SES buying O3b was to reuse the same 3.5-GHz passband centered on 28 GHz. SES has stated that, to date, this has proved to be impractical (personal communication, informal discussion with SES management).

However, putting commercial motivations to one side, we would err towards supporting the OneWeb and Space X and LeoSat claims that cosharing is possible and we base this on the impact of terrestrial antenna innovation, specifically the ability of Flat VSATs to discriminate between wanted energy and unwanted energy.

Note that building a working relationship between the mobile operator community and satellite operators is going to be more achievable if satellite operators are not fighting among themselves, so removing or reducing the inherent spectral tension between NEWLEO, MEO, and GSO operators would be a step forward in itself.

12.3 The Impact of Antenna Innovation on Spectrum Cosharing in Ku-Band, K-Band, and Ka-Band

12.3.1 Active Electronically Steerable Array Antennas (Active Flat VSATs)

This might seem like the board games *Snakes and Ladders* or *Monopoly*, but antenna innovation is crucial to the success or failure of passband cosharing between high-count NEWLEO constellations and MEO and GSO satellites in any band. For the present discussion, we are particularly interested in Ku-band, K-band, and Ka-band for high throughput satellites and in V-band and W-band (E-band) for very high throughput constellations. In Chapter 6, we covered active electronically steerable array (AESA) antennas and their passive equivalents. An AESA can scan from horizon to horizon and actively select the best available satellite connection from a GSO, MEO, or LEO satellite and actively null out signal energy from other satellites. Moving to the K-bands and higher means that compact antennas with 256 elements, 512 elements, or 1,024 elements are entirely practical. Doubling the number of elements realizes 6 dB of gain though the impact of a reduced interference noise floor is potentially more significant. These antennas can be assembled in a repurposed TV LCD display factory or repurposed solar panel factory. The antennas can be shaped to fit the outline of the roof of a car or truck or tank of train or boat or plane. These are called active conformal antennas. Each antenna element has its own RF power amplifier, low noise receive amplifier and filter, filter chain, and phase matching network. This adds cost, particularly as the components need to be specified to work up to +125°C. Operation in high temperatures will also reduce receive sensitivity.

This suggests a cost that would be hard to reduce below a few hundred dollars. While this could be justifiable in a high-end car or truck or train, boat, and plane, it would mean that more price-sensitive markets would be difficult to serve.

12.3.2 Passive Fixed Beamwidth Flat or Conformal Antennas (Passive Flat VSATs)

The alternative is to use passive fixed beamwidth flat or conformal antennas in which the phase offsets of the antenna elements are fixed but arranged such that the antenna looks upwards with a cone of visibility, which could, for example, be 5° or less. These antennas would only see satellites that are directly overhead, for example, a geostationary satellite at the equator or a MEO or high-count LEO at higher latitudes.

This passive option would mean that there could be one RF transceiver for the whole antenna. The transceiver could be remotely mounted thereby avoiding the temperature gradient issues of the active antenna option. Either option would make cosharing of the same passband by LEO, MEO, and GSO constellations more plausible.

12.4 What This Means for the 26 GHz versus 28 GHz Debate

These antenna innovations also make in band cosharing with 5G a more realistic and commercially attractive option. The present "get your tanks off my 28-GHz lawn" is, we would argue, the wrong approach. The satellite industry is lobbying the ITU to make 26 GHz rather than 28 GHz a 5G band despite 28 GHz being chosen as the 5G band in the United States. This robs the satellite industry of any prospect of achieving the scale that it needs to address 5G markets.

By contrast, allowing 5G access to the 28-GHz passband means that economies of scale can be achieved across U.S. and rest of world (ROW) markets and across 5G, 5G point-to-point (PTP), and point-to-multipoint (PTMP) in band backhaul and LEO, MEO, and GSO constellations. The same economies of scale could be achieved by cosharing the 12-GHz, 18-GHz bands and common passbands in the V-band and W-band (E-band).

Combining terrestrial 5G, in-band self-backhaul, and LEO, MEO, and GSO also means that a user or IoT device will always have visibility to multiple connection options. This potentially yields substantial user experience and IoT connectivity benefits. This would represent a major shift of position for the satellite community but could lead to a substantial increase in sector EBITDA and enterprise value. It would also avoid 10 years of litigation and technical argument.

The present position is not dissimilar to the terrestrial TV community claiming that single-frequency networks could not be used for UHF terrestrial broadcasting. Multiple-frequency networks with high protection ratios provided the false technical rationale that 400 MHz of transmission bandwidth was needed to maintain TV broadcast quality. In practice, single-frequency networks have proved efficient and effective and have allowed TV to be packed in to the 500-MHz subband without compromising service quality.

12.5 The Quid Pro Quo: Satellite in the Sub-3.8-GHz 5G Refarming Bands

The satellite industry is not going to countenance any change in access rights or cosharing of K-band, V-band, or E-band spectrum without reciprocal access rights to all bands below 3.8 GHz down to Band 31 5G at 450 MHz. Any change of position would also require agreement on the cosharing of extended C-band (3.8 to 4.2 GHz and above).

12.6 Surely the Satellite Link Budget Is Insufficient for Most Terrestrial Applications?

At this point, you might raise the objection that the path loss for satellites is too high to be useful for many terrestrial applications. Table 12.2 shows the path loss for GEO and LEO satellites assuming a shortest path (vertically down). The path loss will increase with inclination as the path loss will be longer. Rain fade margin also needs to be added in (of the order or 10 dB).

There are several factors to consider. In the event of a rain fade, an AESA antenna will select an alternative satellite with an alternative, hopefully rain-free path. While effective, this will have an impact on the variability of the latency of the end-to-end connection and we have argued that this is an advantage of satellite systems, particularly systems with intersatellite switching. By contrast,

Table 12.2
GEO and LEO Path Loss Comparisons for L-Band and 28 GHz

Path Loss at 28 GHz	
GEO, 212 dB	LEO, 185 dB
Path Loss at L-Band (1.6 GHz)	
GEO, 187 dB	LEO, 152 dB

a passive antenna will always be looking directly upwards and the latency jitter will be less variable though rain will have an impact.

Self-evidently, a user device or IoT device a few meters from a base station with a direct line of sight will have less path loss than a link to a LEO in orbit 700 km away or a MEO 20,000 km away or a GSO 36,000 km away, of the order of a few tens of decibels rather than the larger numbers above. However, the path loss will increase rapidly if the terrestrial path is nonline of sight. Mesh topologies are promoted as a solution but absorb bandwidth and power and increase the local noise floor. As we have stated previously, for outdoor coverage, nearly always nearly overhead or always overhead connectivity from high-count LEO or MEO constellations or GSO satellites over the equator will provide a link that has minimal ground reflection and minimal scatter and minimal surface absorption loss. There will also be 40 to 50 dBi of isotropic gain from the antennas and a significantly lower noise floor. There are therefore many instances in which the link budget from a satellite could be better despite the higher path loss, particularly where terrestrial links are nonline of sight.

12.7 The Satellite Vertical Model

This is best explained by considering the fundamental difference between terrestrial and satellite propagation and path trajectories. We have said that the ideal satellite path trajectory, particularly in the centimeter and millimeter bands, is directly downwards. We have also said that satellite cell footprints can potentially scale from a couple of kilometers to 2,000 km or more (a whole continent for satellite broadcasting, for example). This means that satellites are particularly effective at providing geographic coverage.

12.8 Vertical Coverage for Vertical Markets

Vertical coverage is also particularly effective at providing vertical market coverage. For example, we referenced automotive connectivity and the calculated requirement that an additional 15 to 30 dB of terrestrial link budget would be needed to meet automotive coverage, throughput, latency, and reliability requirements. It will be easier and cheaper to deliver those requirements from a mixed constellation of Ku-band and Ka-band LEO, MEO, and GSO constellations. The same argument can be applied to many other vertical markets including electricity, gas, and water. Note also that we have said that it will only be easier to provide a fully safe autonomous or semiautonomous driving experience from space and more specifically from space constellations with intersatellite and interconstellation switching.

12.9 The Terrestrial Horizontal Mode: Horizontal Coverage for Horizontal Markets

Conversely, terrestrial horizontal coverage is the better option for horizontal markets, for example, outdoor-to-indoor coverage, low-cost, high data rate consumer connectivity, and ultralow-latency local connectivity. Whether satellite or terrestrial is best suited for ultrareliable applications is open to debate, but irrefutably the most reliable link would be one that could access terrestrial 5G and LEO, MEO, and GSO with the option of Ku-band, K-band, Ka-band, and V-band and E-band spectrum cosharing.

12.10 Horizontal versus Vertical Value

5G and satellite are essentially complementary. 5G is best suited to servicing horizontal traffic from 4G and 5G terrestrial base stations and Wi-Fi access points, a combined terrestrial footprint of tens of millions of connectivity points of presence using spectrum from UHF (450-MHz Band 31) to E-band (92 to 95 GHz). Cell sizes scale from a diameter of 20m or less for indoor or highest-density outdoor to 2-km and 20-km cell with larger cells being possible but only efficient at lower frequencies where nonline of sight is not a major problem.

Satellites and high-altitude platforms and helicopters and drone-based 4G or 5G base stations are most efficient when they are servicing traffic directly upwards and directly downwards. Stating the obvious, all these nonterrestrial options are best suited to serving vertical offered traffic.

HAPS and helicopters and drones are good at providing on-demand coverage, for example, in response to a localized emergency. A quasisynchronous self-stabilized HAPS platform at an altitude of 8 to 20 km could potentially provide cost-effective coverage to a 200-km cell, although with relatively low elevation angles (10%) at the cell edge. These platforms are also effective spies in the sky and can perform a range of imaging and sensing and incident monitoring functions.

There are significant advantages in getting above the clouds and into space particularly as satellites are now lasting 20 years or more, providing a long-term capital amortization opportunity with potentially low operational costs including free rent and electricity. It may be that a significant percentage of traffic presently serviced horizontally from terrestrial networks could be serviced more efficiently from satellites, particularly satellites that are directly overhead. Note that LEO, MEO, and GSO satellites can also deliver traffic horizontally using intersatellite switching. This can be faster and more efficient than delivering traffic over a terrestrial fiber, cable, or copper network. Last but not least, we can route traffic upwards using inter constellation switching with LEO satellites sending traffic upwards to MEO and GSO satellites and then back to Earth through existing fully amortized GSO ground stations.

Innovations in launch and satellite technology now make it feasible to launch thousands rather than hundreds of satellites into LEO and MEO and ever larger and more powerful satellites into GSO.

Although it seems presently unlikely that terrestrial base station density could be replicated in near space, it is not impossible provided issues such as space debris could be managed effectively and efficiently. The energy savings over 20 years could outweigh the carbon cost of the initial launch, so it is possible that space-based networks could be a lower carbon connectivity option, although wind and solar-powered terrestrial networks would also need to be factored into this calculation.

Antenna innovation on satellites allows cell diameters on Earth to be scaled from 2 km–20 km to 200 km–2,000 km or more and use spectrum from VHF to E-band. However, this rather misses the point. 5G and satellite systems including high throughput satellites in Ku-band and Ka-band, very high throughput satellites in V-band and W-band, and superhigh throughput satellites in E-band are all beam-based networks. They are effectively providing progressive point-to-point coverage for individual users or small groups of users with beam-to-beam handover.

For satellites, it is better to think spatially and not think about a cell but of the cone of visibility provided by the nearest satellite passing overhead with handover being performed between satellites and traffic being routed through the intersatellite and interconstellation switching matrix. In the frequency domain, there could be separation between local and backhaul traffic and separation between the user plane or the control plane or it could all be multiplexed together in 250-MHz or 500-MHz channels within a 3.5 GHz + 3.5 GHz or 5 GHz + 5 GHz passband.

Either way, 5G terrestrial and satellite networks use steerable beams or switchable fixed beams to replicate guided media performance by narrowing beam patterns to the point at which only the RF energy of interest gets received into the antenna on the receive path at both ends of the link. Conversely, unwanted signal energy is kept out of spectrally and geographically proximate systems.

These are all essentially progressive point-to-point systems although with a make-before-break, beam-to-beam handover to maintain individual traffic flows. The most optimum operating point in terms of link efficiency is directly overhead, although this implies high-count constellations (hundreds or thousands of satellites) for LEO, tens or hundreds of satellites for MEO, and ideally 40 or more GSO satellites evenly distributed around the equators (to minimize east to west elevation).

The concept of minimizing available elevation is awkward commercially as it implies that GSO satellites would be constrained from servicing users at higher northern and southern latitudes. Also, it is generally only practical to provide in building coverage from low elevation angles (unless roof-mounted antennas are used).

Nevertheless, if a GSO operator also owns or has access to MEO and high-count LEO bandwidth, then it would make sense to service users and IoT devices from the satellite with the shortest link, which will always be from directly overhead or as nearly directly overhead as practical and will be mostly line of sight. Not only does this minimize latency and maximize the link budget, but it also avoids the surface absorption and scatter and ground reflections, which will be problematic for terrestrial networks particularly in areas with limited line-of-sight access.

If a LEO, MEO, and GSO all happen to be directly overhead at the same time, then all three systems will have to be coordinated, but in principle could be combined to maximize flux density. From a business modeling point of view, this suggests a need to establish the cost of providing vertical bandwidth from space and the equivalent cost of providing the same coverage and service from a terrestrial network in the horizontal plane and then to quantify the value that can be realized from both options either singly or together.

As stated in Chapter 1, there are some applications where satellites are the only option (maritime and deep rural, for example), some applications where satellites are more efficient and effective and some applications where 5G and Wi-Fi are more efficient and effective with satellites potentially having a broader role than presently envisaged.

12.11 Summary: Around the World in 80 Ways

Skybridge and Teledesic introduced the concept of high-count LEO constellations that could connect the world cost-effectively by using in orbit progressive pitch mechanisms to enable spectrum cosharing. This was 20 years ago.

In Europe, a similar proposal was proposed that came to be known as the Eighty LEO constellation [1].

Teledesic and Skybridge no longer exist, although we have documented that many of the techniques proposed have formed the basis of the present constellations being implemented by U.S. companies including OneWeb and Space X. There are probably at least 80 ways we can send and receive voice and data round the world and some ways are better (more efficient and effective) than others.

The satellite industry is undergoing a remarkable technical transformation, which includes launch innovation, satellite and constellation innovation, and production and manufacturing innovation. The satellite industry is also undergoing a remarkable commercial transformation with a new generation of entrepreneurs backed by Web-scale companies such as Google, Facebook, Amazon, Alibaba, and Tencent. These Web-scale companies have cash, customers, and the algorithms needed to extract value from large data sets. Satellites are particularly well suited to data acquisition particularly when the satellite service offer is closely coupled to a discrete vertical market.

In the automotive industry, Pirelli are capturing sensor data including tire data from cars and uploading the data to the Pirelli cloud for resale or to improve commercial efficiency, for example, by telling customers when their tires need replacing or their tracking needs attention. This can be done over the cellular network but arguably can be done more effectively from a mixed satellite constellation. The Doppler signature of MEO and LEO satellites also provides an alternative positioning mechanism to a potential accuracy of a nanosecond (1 foot of location ambiguity). Combining these mechanisms with Quazi zenith constellations provides even more robust positioning and location. Satellites can also amortize costs across communication, imaging, and sensing and across commercial and military payloads.

However, the satellite industry does not have 4 billion smartphone customers and lacks the standards bandwidth, which has been key to the success of the mobile and fixed wireless broadband revolution. The satellite industry has a supply chain optimized for producing a few hundred exquisitely engineered satellites rather than millions of base stations and access points. Smartphone designs and the materials and manufacturing innovation that brings these designs to market at consumer price points are the product of scale. Google with 1 billion users and Facebook with 2 billion users do not have this manufacturing and materials capability and, to date, have been unsuccessful at buying into a complex and brutally efficient mobile broadband device and network vendor supply chain, which invests 12% to 14% of its turnover on research and development.

Conversely, the 5G industry has problems to solve and the satellite industry can help solve them. This includes an escalating cost base due to network densification, a legacy investment focus that has prioritized demographic over geographic coverage, an energy efficiency problem (also the product of network densification), and a carbon footprint problem.

The satellite industry can deliver tens of satellites into space on a single rocket, arriving in space a few minutes after takeoff; the satellites can stay in space for 20 years, pay no ground rent, and have a limitless source of free electricity.

The technical opportunity exists for LEO, MEO, and GSO operators to coshare their spectrum with the 5G industry. This would avoid 10 years of essentially fruitless technical dispute and litigation but more importantly would transform the delivery and scale economics of the whole industry and transform the user experience across consumer and vertical markets and the economics of terrestrial and maritime IoT.

The biggest prize of all is to add satellite connectivity to 5G smartphones and wearable devices. The additional coverage footprint would improve the EBITDA and enterprise value of the mobile operators. Satellite operators would gain by realizing access to connected consumer added value making them equal partners rather than supplicants to the ever more dominant over the top Web-scale corporate sector. In a dog-eats-dog world, the chihuahua [2] needs to outwit the pit bull terrier [3].

One hundred years ago, the first wireless telegraphy message was exchanged between the United Kingdom and Australia, masterminded by Marconi, the man who modestly described himself as the man who "connected the world by wireless." Today, we are starting on a second century of global wireless innovation. It is just possible that Mr. Musk, in many ways a modern Marconi, may similarly be remembered 100 years from now as the man who connected the world from space, but who will be his fellow travelers?

References

[1] http://www.eightyleo.com/.

[2] http://www.chiwawadog.com/.

[3] http://www.bbc.co.uk/newsbeat/article/36367983/the-dog-breeds-that-are-banned-in-the-uk-and-why.

About the Author

Geoff Varrall joined RTT in 1985 as an executive director and shareholder to develop RTT's international business as a provider of technology and business services to the wireless industry.

He codeveloped RTT's original series of design and facilitation workshops, including *RF Technology*, *Data over Radio*, *Introduction to Mobile Radio*, and *Private Mobile Radio Systems*, and also developed *The Oxford Programme*, a five day strategic technology and market program presented annually between 1991 and 2005. Geoff has been running in-depth technology and market workshops for the industry for over 33 years, spanning five generations of mobile, cellular, and Wi-Fi technologies.

Mr. Varrall is a coauthor of the *Mobile Radio Servicing Handbook* (Heinemann Butterworth, UK), *Data Over Radio* (Quantum Publishing, Mendocino, USA) and *3G Handset and Network Design* (John Wiley, New York).

Mr. Varrall's fourth book, *Making Telecoms Work—From Technical Innovation to Commercial Success* (John Wiley) was published in early 2012. His fifth book, *5G Spectrum and Standards* was published by Artech House in July 2016. This book, *5G and Satellite Spectrum, Standards and Scale* brings the 5G story up to date and places the 5G business model into a broader connectivity context, including satellite and Wi-Fi products, networks, and systems.

Mr. Varrall regularly presents workshops and master classes in the UK, Singapore, Southeast Asia, and Australasia.

In his spare time he plays jazz trumpet, flügelhorn, and cornet semiprofessionally, and is a keen marathon runner and ultrarunner.

Index

Artech House
Space Technology and Applications Series

Bruce R. Elbert, Series Editor

5G and Satellite Spectrum, Standards, and Scale, Geoff Varrall

Business Strategies for Satellite Systems, D. K. Sachdev

Delay-Tolerant Satellite Networks, Juan A. Fraire, Jorge M. Finochietto, and Scott C. Burleigh

Filter Design for Satellite Communications: Helical Resonator Technology, Efstratios Doumanis, George Goussetis, and Savvas Kosmopoulos

Gigahertz and Terahertz Technologies for Broadband Communications, Terry Edwards

Ground Segment and Earth Station Handbook, Bruce R. Elbert

Introduction to Satellite Communication, Third Edition, Bruce R. Elbert

Laser Space Communications, David G. Aviv

Low Earth Orbital Satellites for Personal Communication Networks, Abbas Jamalipour

Mobile Satellite Communications, Shingo Ohmori, et al.

Radio Frequency Inteference in Communications Systems, Bruce R. Elbert

Radio Interferometry and Satellite Tracking, Seiichiro Kawase

Satellite Broadcast Systems Engineering, Jorge Matos Gómez

The Satellite Communication Applications Handbook, *Second Edition*, Bruce R. Elbert

Satellite Communications Fundamentals, Jules E. Kadish and Thomas W. R. East

Satellite Communications Network Design and Analysis, Kenneth Y. Jo

The Satellite Communication Ground Segment and Earth Station Handbook, Second Edition, Bruce Elbert

Space Microelectronics, Volume 1, Modern Spacecraft Classification, Failure, and Electrical Component Requirements, Anatoly Belous, Vitali Saladukha, and Siarhei Shvedau

Space Microelectronics, Volume 2, Integrated Circuit Design for Space Applications, Anatoly Belous, Vitali Saladukha, and Siarhei Shvedau

Understanding GPS: Principles and Applications, Second Edition,
Elliott D. Kaplan and Christopher J. Hegarty, editors

For further information on these and other Artech House titles, including out-of-print books available through our In-Print-Forever® (IPF®) program, contact:

Artech House
685 Canton Street
Norwood, MA 02062
Phone: 781-769-9750
Fax: 781-769-6334
e-mail: artech@artechhouse.com

Artech House
16 Sussex Street
London SW1V 4RW U.K.
Phone: +44 (0)171-973-8077
Fax: +44 (0)171-630-0166
e-mail: artech-uk@artechhouse.com

Find us on the World Wide Web at: www.artechhouse.com